2006

STRUCTURAL/SEISMIC DESIGN MANUAL

1

CODE APPLICATION EXAMPLES

Publisher

Structural Engineers Association of California (SEAOC)

1414 K Street, Suite 260
Sacramento, California 95814
Telephone: (916) 447-1198; Fax: (916) 443-8065
E-mail: lee@seaoc.org; Web address: www.seaoc.org

The Structural Engineers Association of California (SEAOC) is a professional association of four regional member organizations (Southern California, Northern California, San Diego, and Central California). SEAOC represents the structural engineering community in California. This document is published in keeping with SEAOC's stated mission: "to advance the structural engineering profession; to provide the public with structures of dependable performance through the application of state-of-the-art structural engineering principles; to assist the public in obtaining professional structural engineering services; to promote natural hazard mitigation; to provide continuing education and encourage research; to provide structural engineers with the most current information and tools to improve their practice; and to maintain the honor and dignity of the profession."

Editor

International Code Council

Second Printing: July 2007
Third Printing: October 2007
Fourth Printing: January 2008

Table of Contents

Preface

This document is the initial volume in the three-volume 2006 *IBC Structural/Seismic Design Manual*. It has been developed by the Structural Engineers Association of California (SEAOC) with funding provided by SEAOC. Its purpose is to provide guidance on the interpretation and use of the seismic requirements in the 2006 *International Building Code* (IBC), published by the International Code Council, Inc., and SEAOC's 2005 *Recommended Lateral Force Requirements and Commentary* (also called the Blue Book).

The 2006 *IBC Structural/Seismic Design Manual* was developed to fill a void that exists between the commentary of the Blue Book, which explains the basis for the code provisions, and everyday structural engineering design practice. The 2006 *IBC Structural/Seismic Design Manual* illustrates how the provisions of the code are used. *Volume I: Code Application Examples,* provides step-by-step examples for using individual code provisions, such as computing base shear or building period. *Volumes II and III: Building Design Examples,* furnish examples of seismic design of common types of buildings. In Volumes II and III, important aspects of whole buildings are designed to show, calculation-by-calculation, how the various seismic requirements of the code are implemented in a realistic design.

The examples in the 2006 *IBC Structural/Seismic Design Manual* do not necessarily illustrate the only appropriate methods of design and analysis. Proper engineering judgment should always be exercised when applying these examples to real projects. The 2006 *IBC Structural/Seismic Design Manual* is not meant to establish a minimum standard of care but, instead, presents reasonable approaches to solving problems typically encountered in structural / seismic design.

The example problem numbers used in the prior Seismic Design Manual – 2000 IBC Volume I code application problems have been retained herein to provide easy reference to compare revised code requirements.

SEAOC, NCSEA and ICC intend to update the 2006 *IBC Structural/Seismic Design Manual* with each edition of the building code.

Jon P. Kiland and Rafael Sabelli
Project Managers

Acknowledgments

Authors

The 2006 *IBC Structural/Seismic Design Manual Volume 1* was written by a group of highly qualified structural engineers. They were selected by a steering committee set up by the SEAOC Board of Directors and were chosen for their knowledge and experience with structural engineering practice and seismic design. The consultants for Volumes I, II, and III are:

Jon P. Kiland, *Co-Project Manager*
Rafael Sabelli, *Co-Project Manager*
Douglas S. Thompson
Dan Werdowatz
Matt Eatherton

John W. Lawson
Joe Maffei
Kevin Moore
Stephen Kerr

Reviewers

A number of SEAOC members and other structural engineers helped check the examples in this volume. During its development, drafts of the examples were sent to these individuals. Their help was sought in review of code interpretations as well as detailed checking of the numerical computations.

Seismology Committee

Close collaboration with the SEAOC Seismology Committee was maintained during the development of the document. The 2004-2005 and 2005-2006 committees reviewed the document and provided many helpful comments and suggestions. Their assistance is gratefully acknowledged.

Production and Art

ICC

Suggestions for Improvement

In keeping with SEAOC's and NCSEA's Mission Statements: "to advance the structural engineering profession" and "to provide structural engineers with the most current information and tools to improve their practice," SEAOC and NCSEA plan to update this document as structural/seismic requirements change and new research and better understanding of building performance in earthquakes becomes available.

Comments and suggestions for improvements are welcome and should be sent to the following:

Structural Engineers Association of California (SEAOC)
Attention: Executive Director
1414 K Street, Suite 260
Sacramento, California 95814
Telephone: (916) 447-1198; Fax: (916) 932-2209
E-mail: lee@seaoc.org; Web address: www.seaoc.org

Errata Notification

SEAOC and NCSEA have made a substantial effort to ensure that the information in this document is accurate. In the event that corrections or clarifications are needed, these will be posted on the SEAOC web site at *http://www.seaoc.org* or on the ICC website at *http://www.iccsafe.org*. SEAOC, at its sole discretion, may or may not issue written errata

Introduction

Volume I of the 2006 *IBC Structural/Seismic Design Manual: Code Application Examples* deals with interpretation and use of the structural/seismic provisions of the 2006 *International Building Code®* (IBC). The 2006 *IBC Structural/Seismic Design Manual* is intended to help the reader understand and correctly use the IBC structural/seismic provisions and to provide clear, concise, and graphic guidance on the application of specific provisions of the code. It primarily addresses the major structural/seismic provisions of the IBC, with interpretation of specific provisions and examples highlighting their proper application.

The 2006 IBC has had structural provisions removed from its text and has referenced several national standards documents for structural design provisions. The primary referenced document is ASCE/SEI 7-05, which contains the "Minimum Design Loads for Buildings and Other Structures." ASCE/SEI 7-05 is referenced for load and deformation design demands on structural elements, National Material design standards (such as ACI, AISC, MSJC and NDS) are then referenced to take the structural load demands from ASCE/SEI 7-05 and perform specific material designs.

Volume I presents 58 examples that illustrate the application of specific structural/seismic provisions of the IBC. Each example is a separate problem, or group of problems, and deals primarily with a single code provision. Each example begins with a description of the problem to be solved and a statement of given information. The problem is solved through the normal sequence of steps, each of which is illustrated in full. Appropriate code references for each step are identified in the right-hand margin of the page.

The complete 2006 *IBC Structural/Seismic Design Manual* will have three volumes. Volumes II and III will provide a series of structural/seismic design examples for buildings illustrating the seismic design of key parts of common building types such as a large three-story wood frame building, a tilt-up warehouse, a braced steel frame building, and a concrete shear wall building.

While the 2006 *IBC Structural/Seismic Design Manual* is based on the 2006 IBC, there are some provision of SEAOC's 2005 *Recommended Lateral Force Provisions and Commentary* (Blue Book) that are applicable. When differences between the IBC and Blue Book are significant they are brought to the attention of the reader.

The 2006 *IBC Structural/Seismic Design Manual* is intended for use by practicing structural engineers and structural designers, building departments, other plan review agencies, and structural engineering students.

How to Use This Document

The various code application examples of Volume I are organized by topic consistent with previous editions. To find an example for a particular provision of the code, look at the upper, outer corner of each page, or in the table of contents.

Generally, the ASCE/SEI 7-05 notation is used throughout. Some other notation is defined in the following pages, or in the examples.

Reference to ASCE/SEI 7-05 sections and formulas is abbreviated. For example, "ASCE/SEI 7-05 §6.4.2" is given as §6.4.2 with ASCE/SEI 7-05 being understood. "Equation (12.8-3)" is designated (Eq 12.8-3) in the right-hand margins. Similarly, the phrase "T 12.3-1" is understood to be ASCE/SEI 7-05 Table 12.3-1, and "F 22-15" is understood to be Figure 22-15. Throughout the document, reference to specific code provisions and equations is given in the right-hand margin under the category Code Reference.

Generally, the examples are presented in the following format. First, there is a statement of the example to be solved, including given information, diagrams, and sketches. This is followed by the "Calculations and Discussion" section, which provides the solution to the example and appropriate discussion to assist the reader. Finally, many of the examples have a third section designated "Commentary." In this section, comments and discussion on the example and related material are made. Commentary is intended to provide a better understanding of the example and/or to offer guidance to the reader on use of the information generated in the example.

In general, the Volume I examples focus entirely on use of specific provisions of the code. No building design is illustrated. Building design examples are given in Volumes II and III.

The 2006 *IBC Structural/Seismic Design Manual* is based on the 2006 IBC, and the referenced Standard ASCE/SEI 7-05 unless otherwise indicated. Occasionally, reference is made to other codes and standards (e.g., 2005 AISC Steel Construction Manual 13th Edition, ACI 318-05, or 2005 NDS). When this is done, these documents are clearly identified.

Notation

The following notations are used in this document. These are generally consistent with those used in ASCE/SEI 7-05 and other Standards such as ACI and AISC. Some new notations have also been added. The reader is cautioned that the same notation may be used more than once and may carry entirely different meanings in different situations. For example, E can mean the tabulated elastic modulus under the AISC definition (steel) or it can mean the earthquake load under §12.4.2 of ASCE/SEI 7-05. When the same notation is used in two or more definitions, each definition is prefaced with a brief description in parentheses (e.g., steel or loads) before the definition is given.

A = area of floor or roof supported by a member

A_{BM} = cross-sectional area of the base material

A_b = area of anchor, in square inches

A_c = the combined effective area, in square feet, of the shear walls in the first story of the structure

A_{ch} = cross-sectional area of a structural member measured out-to-out of transverse reinforcement

A_{cv} = net area of concrete section bounded by web thickness and length of section in the direction of shear force considered

A_e = the minimum cross-sectional area in any horizontal plane in the first story, in square feet of a shear wall

A_f = flange area

A_g = gross area of section

A_g = the gross area of that wall in which A_o is identified

A_i = the floor area in square feet of the diaphragm level immediately above the story under consideration

A_o = area of the load-carrying foundation

A_p = the effective area of the projection of an assumed concrete failure surface upon the surface from which the anchor protrudes, in square inches

A_s = area of non-prestressed tension reinforcement

A_{sh}	=	total cross-sectional area of transverse reinforcement (including supplementary crossties) having a spacing s_n and crossing a section with a core dimension of h_c
A_{sk}	=	area of skin reinforcement per unit height in one side face
$A_{s,min}$	=	area having minimum amount of flexural reinforcement
A_{st}	=	area of link stiffener
A_T	=	tributary area
A_v	=	area of shear reinforcement within a distance s, or area of shear reinforcement perpendicular to flexural tension reinforcement within a distance s for deep flexural members
A_{vd}	=	required area of leg reinforcement in each group of diagonal bars in a diagonally reinforced coupling beam
A_{vf}	=	area of shear-friction reinforcement
A_w	=	(web) link web area
A_w	=	(weld) effective cross-sectional area of the weld
A_x	=	the torsional amplification factor at Level x – §12.8.4.3
a	=	(concrete) depth of equivalent rectangular stress block
a	=	(concrete spandrel) shear span, distance between concentrated load and face of supports
a_c	=	coefficient defining the relative contribution of concrete strength to wall strength
a_d	=	incremental factor relating to the P-delta effects as determined in §12.8.7
a_i	=	the acceleration at Level i obtained from a modal analysis (§13.3.1)
a_p	=	amplification factor related to the response of a system or component as affected by the type of seismic attachment determined in §13.3.1
b	=	(concrete) width of compression face of member
b_f	=	flange width
b_w	=	web width

b/t	$=$	member width-thickness ratio
C_d	$=$	deflection amplification factor as given in Tables 12.2-1 or 15.4-1 or 15.4-2
C_e	$=$	snow exposure factor
C_m	$=$	coefficient defined in §H1 of AISC/ASD, 9th Edition
C_s	$=$	the seismic response coefficient determined in §12.8.1.1 and §19.3.1
C_T	$=$	building period coefficient – §12.8.2.1
C_t	$=$	snow thermal factor
C_{vx}	$=$	vertical distribution factor – §12.8.3
c	$=$	distance from extreme compression fiber to neutral axis of a flexural member
D	$=$	dead load, the effect of dead load
D_e	$=$	the length, in feet, of a shear wall in the first story in the direction parallel to the applied forces
D_h	$=$	gross weight of helicopter
D_p	$=$	relative seismic displacement that a component must be designed to accommodate – §13.3.2
d	$=$	effective depth of section (distance from extreme compression fiber to centroid of tension reinforcement)
d_b	$=$	(anchor bolt) anchor shank diameter
d_b	$=$	(concrete) bar diameter
d_z	$=$	column panel zone depth
E	$=$	(steel) modulus of elasticity
E	$=$	combined effect of horizontal and vertical earthquake-induced forces (§12.4)
E_m	$=$	seismic load effect including overstrength factors (§§12.4.3.2 and 12.14.2.2.2)

EI	$=$	flexural stiffness of compression member
E_c	$=$	modules of elasticity of concrete, in psi
E_s	$=$	(concrete) modulus of elasticity of reinforcement
e	$=$	EBF link length
F	$=$	load due to fluids
F_a	$=$	site coefficient defined in §11.4.3
F_a	$=$	axial compressive stress that would be permitted if axial force alone existed
F_a	$=$	flood load
F_b	$=$	bending stress that would be permitted if bending moment alone existed
F_{BM}	$=$	nominal strength of the base material to be welded
F_{EXX}	$=$	classification number of weld metal (minimum specified strength)
F_i, F_n, F_x	$=$	portion of seismic base shear, V, induced at Level i, n, or x as determined in §12.8.3.
F_p	$=$	seismic force, induced by the parts being connected, centered at the component's center of gravity and distributed relative to the component's mass distribution, as determined in §12.8.3
F_{px}	$=$	the diaphragm design force
F_u	$=$	specified minimum tensile strength, ksi
F_u	$=$	through-thickness weld stresses at the beam-column interface
F_{ut}	$=$	minimum specified tensile strength of the anchor
F_v	$=$	long period site coefficient (at 1.0 second period) see §11.4.3
F_x	$=$	the design lateral force applied at Level x
F_x	$=$	the lateral force induced at any Level i – §12.8.3
F_w	$=$	(steel LRFD) nominal strength of the weld electrode material
F_w	$=$	(steel ASD) allowable weld stress

F_y	=	specified yield strength of structural steel
F_{yb}	=	F_y of a beam
F_{yc}	=	F_y of a column
F_{ye}	=	expected yield strength of steel to be used
F_{yf}	=	F_y of column flange
F_{yh}	=	(steel) specified minimum yield strength of transverse reinforcement
F_{yw}	=	F_y of the panel-zone steel
f_1	=	load factor applied to live load in load combinations – IBC §1605.2
f_2	=	snow load reduction factor – IBC §1605
f_a	=	computed axial stress
f_b	=	bending stress in frame member
f'_c	=	specified compressive strength of concrete
f_{ct}	=	average splitting tensile strength of lightweight aggregate concrete
f'_m	=	specified compressive strength of masonry
f_p	=	equivalent uniform load
f_r	=	modulus of rupture of concrete
f_y	=	(concrete) specified yield strength of reinforcing steel
f_{yn}	=	(concrete) specified yield strength of special lateral reinforcement
g	=	acceleration due to gravity (gravitational acceleration constant 32.2 ft/sec^2 or 386.4 in/sec^2)
H	=	load due to lateral pressure of soil and water in soil
h	=	average roof height of structure relative to the base elevation
h	=	overall dimensions of member in direction of action considered
h_c	=	(concrete) cross-sectional dimension of column core, or shear wall boundary zone, measured center-to-center of confining reinforcement
h_c	=	(steel) assumed web depth for stability

h_e	=	assumed web depth for stability
h_i, h_n, h_x	=	height in feet above the base to Level i, n or x, respectively
h_r	=	height in feet of the roof above the base
h_{sx}	=	the story height below Level x
h_w	=	height of entire wall or of the segment of wall considered
I	=	the importance factor determined in accordance with §11.5.1
I	=	moment of inertia of section resisting externally applied factored loads
I_{cr}	=	moment of inertia of cracked section transformed to concrete
I_g	=	(concrete, neglecting reinforcement) moment of inertia of gross concrete section about centroidal axis
I_{se}	=	moment of inertia of reinforcement about centroidal axis of member cross section.
I_t	=	moment of inertia of structural steel shape, pipe or tubing about centroidal axis of composite member cross section.
I_g	=	(concrete, neglecting reinforcement) moment of inertia of gross concrete section about centroidal axis, neglecting reinforcement.
I_p	=	component importance factor that is either 1.00 or 1.5, as determined in §13.3.1
K	=	(steel) effective length factor for prismatic member
k	=	a distribution exponent – §12.8.3
L	=	live load, except roof live load, including any permitted live load reduction (i.e. reduced design live load). Live load related internal moments or forces. Concentrated impact loads
L_o	=	unreduced design live load
L_b	=	(steel) unbraced beam length for determining allowable bending stress
L_p	=	limiting laterally unbraced length for full plastic flexural strength, uniform moment case
L_r	=	roof live load including any permitted live load reduction

l_c	=	(steel RBS) length of radius cut in beam flange for reduced beam section (RBS) design
l_c	=	length of a compression member in a frame, measured from center to center of the joints in the frame
l_h	=	distance from column centerline to centerline of hinge for reduced bending strength (RBS) connection design
l_n	=	clear span measured face-to-face of supports
l_u	=	unsupported length of compression member
l_w	=	length of entire wall, or of segment of wall considered, in direction of shear force
Level i	=	level of the structure referred to by the subscript i. "$i = 1$" designates the first level above the base
Level n	=	that level that is uppermost in the main portion of the structure
Level x	=	that level that is under design consideration. "$x = 1$" designates the first level above the base
M	=	(steel) maximum factored moment
M_c	=	factored moment to be used for design of compression member
M_{cl}	=	moment at centerline of column
M_{cr}	=	moment at which flexural cracking occurs in response to externally applied loads
M_{DL}, M_{LL}, M_{seis}	=	limiting laterally unbraced length for full plastic flexural strength, uniform moment case
M_f	=	moment at face of column
M_m	=	(concrete) modified moment
M_m	=	(steel) maximum moment that can be resisted by the member in the absence of axial load
M_n	=	(steel) nominal moment strength at section
M_p	=	(concrete) required plastic moment strength of shearhead cross section

M_p	=	(steel) nominal plastic flexural strength, $F_y Z$
M_{pa}	=	nominal plastic flexural strength modified by axial load
M_{pe}	=	nominal plastic flexural strength using expected yield strength of steel
M_{pr}	=	(concrete) probable moment strength determined using a tensile strength in the longitudinal bars of at least $1.25 f_y$ and a strength reduction factor ϕ of 1.0
M_{pr}	=	(steel RBS) probable plastic moment at the reduced beam section (RBS)
M_s	=	(concrete) moment due to loads causing appreciable sway
M_t	=	torsional moment
M_{ta}	=	accidental torsional moment
M_u	=	(concrete) factored moment at section
M_u	=	(steel) required flexural strength on a member or joint
M_y	=	moment corresponding to onset of yielding at the extreme fiber from an elastic stress distribution
M_1	=	smaller factored end moment on a compression member, positive if member is bent in single curvature, negative if bent in double curvature
M_2	=	larger factored end moment on compression member, always positive
N	=	number of stories
P	=	ponding load
P	=	(steel) factored axial load
P	=	(wind) design wind pressure
P_{DL}, P_{LL}, P_{seis}	=	unfactored axial load in frame member
P_b	=	nominal axial load strength at balanced strain conditions
P_{bf}	=	connection force for design of column continuity plates
P_c	=	(concrete) critical load

P_c	=	(concrete anchorage) design tensile strength
P_n	=	nominal axial load strength at given eccentricity, or nominal axial strength of a column
P_o	=	nominal axial load strength at zero eccentricity
P_{si}	=	$F_y A$
P_u	=	(concrete) factored axial load, or factored axial load at given eccentricity
P_u	=	(steel) nominal axial strength of a column, or required axial strength on a column or a link
P_u	=	(concrete anchorage) required tensile strength from loads
P_y	=	nominal axial yield strength of a member, which is equal to $F_y A_g$
P_x	=	total unfactored vertical design load at and above Level x
P_E	=	axial load on member due to earthquake
P_{LL}	=	axial live load
Q_E	=	the effect of horizontal seismic forces
R	=	rain load
R	=	The response modification factor from Table 12.2-1
R_n	=	nominal strength
R_p	=	component response modification factor that varies from 1.00 to 3.50 as set forth in Table 13.5-1 or Table 13.6-1
R_u	=	required strength
R_y	=	ratio of expected yield strength F_{ye} to the minimum specified yield strength F_y
$R, R_1 R_2$	=	live load reduction in percent – IBC §§1607.9.2/1607.11.2
r	=	rate of reduction equal to 0.08 percent for floors
r	=	(steel) radius of gyration of cross section of a compression member
r_y	=	radius of gyration about y axis

S	=	snow load
S_a	=	design spectral response acceleration
	=	$0.6\,(S_{DS}/T_o)\,T + 0.4\,(S_{DS})$, for T less than or equal to T_o
	=	$(S_{D1})\,/\,T$, for T greater than T_s
S_{DS}	=	5% damped, design, spectral response acceleration parameter at short period (i.e., 0.2 seconds) = $(2/3)\,S_{ms}$ – §11.4.4
S_s	=	Mapped, MCE, 5% damped, spectral acceleration parameter at short periods (i.e., 0.2 seconds) as determined by §11.4.1
S_{D1}	=	5% damped, design, spectral response acceleration parameter at 1-second period = $(2/3)\,S_{M1}$
S_1	=	Mapped, MCE, 5% damped, spectral acceleration parameter for a 1-second period as determined in §11.4.1
S_{MS}	=	MCE, 5% damped, spectral response acceleration parameter for short periods (i.e., 0.2 seconds) = $F_a S_s$, adjusted for site class effects
S_{M1}	=	MCE, 5% damped, spectral response acceleration parameter for 1-second period = $F_v S_1$, adjusted for site class effects
S_{RBS}	=	section modulus at the reduced beam section (RBS)
s	=	spacing of shear or torsion reinforcement in direction parallel to longitudinal reinforcement, or spacing of transverse reinforcement measured along the longitudinal axis
T	=	self-straining force arising from contraction or expansion resulting from temperature change, shrinkage, moisture change, creep in component materials, movement due to differential settlement or combinations thereof
T	=	elastic fundamental period of vibration, in seconds, of the structure in the direction under consideration, see §11.4.5 for limitations
T_a	=	approximate fundamental period as determined in accordance with §12.8.2.1
T_o	=	$0.2\,(S_{D1}\,/\,S_{DS})$
T_s	=	$S_{D1}\,/\,S_{DS}$
t_f	=	thickness of flange
t_w	=	thickness of web

t_z	=	ratio of expected yield strength F_{ye} to the minimum specified yield strength F_y
U	=	required strength to resist factored loads or related internal moments and forces
V	=	the total design seismic lateral force or shear at the base of the building or structure
V_c	=	(concrete) nominal shear strength provided by concrete
V_c	=	(concrete anchorage) design shear strength
V_{DL}, V_{LL}, V_{seis}	=	unfactored shear in frame member
V_m	=	shear corresponding to the development of the "nominal flexural strength – calculated in accordance with Chapter 19"
V_n	=	(concrete) nominal shear strength at section
V_n	=	(steel) nominal shear strength of a member
V_p	=	(steel) shear strength of an active link
V_{pa}	=	nominal shear strength of an active link modified by the axial load magnitude
V_{px}	=	the portion of the seismic shear force at the level of the diaphragm, required to be transferred to the components of the vertical seismic-lateral-force-resisting system because of the offsets or changes in stiffness of the components above or below the diaphragm
V_s	=	(concrete) nominal shear strength provided by shear reinforcement
V_s	=	(steel) shear strength of member, $0.55 F_y dt$
V_u	=	(concrete anchorage) required shear strength from factored loads
V_u	=	(concrete) factored shear force at section
V_u	=	(loads) factored horizontal shear in a story
V_u	=	(steel) required shear strength on a member
V_x	=	the seismic design story shear (force) in story x, (i.e., between Level x and x-1)

W = the total effective seismic dead load (weight) defined in §12.7.2 and §12.14.8.1

W = (wind) load due to wind pressure

W_p = component operating weight

w_c = weights of concrete, in pcf

w_i, w_x = that portion of W located at or assigned to Level i or x, respectively

w_p = the weight of the smaller portion of the structure

w_p = the weight of the diaphragm and other elements of the structure tributary to the diaphragm

w_{px} = the weight of the diaphragm and elements tributary thereto at Level x, including applicable portions of other loads defined in §12.7.2

w_w = weight of the wall tributary to the anchor

w_z = column panel zone width

X = height of upper support attachment at Level x as measured from the base

Y = height of lower support attachment at Level Y as measured from the base

Z = (steel) plastic section modulus

z = height in structure at point of attachment of component, §13.3.1

Z_{RBS} = plastic section modulus at the reduced beam section (RBS)

ϕ = (concrete) capacity-reduction or strength-reduction factor

ϕ_b = (steel) resistance factor for flexure

ϕ_c = (steel) resistance factor for compression

ϕ_V = resistance factor for shear strength of panel-zone of beam-to-column connections

α = (concrete) angle between the diagonal reinforcement and the longitudinal axis of a diagonally reinforced coupling beam

α, β =   (steel) centroid locations of gusset connection for braced frame diagonal

α_c =   coefficient defining the relative contribution of concrete strength to wall strength

β_c =   ratio of long side to short side of concentrated load or reaction area

β =   the ratio of shear demand to shear capacity for the story between Level x and x-1

ρ =   a redundancy factor determined in accordance with §12.3.4

ρ =   (concrete) ratio of nonprestressed tension reinforcement (A_s/b_d)

ρ_b =   reinforcement ratio producing balanced strain conditions

ρ_n =   ratio of area of distributed reinforcement parallel to the plane of A_{cv} to gross concrete area perpendicular to the reinforcement

ρ_s =   ratio of volume of spiral reinforcement to total volume of core (out-to-out of spirals) of a spirally reinforced compression member

ρ_v =   ratio of area of distributed reinforcement perpendicular to the plane of A_{cv} to gross concrete area A_{cv}

λ =   lightweight aggregate concrete factor; 1.0 for normal-weight concrete, 0.75 for "all lightweight" concrete, and 0.85 for "sand-lightweight" concrete

λ_p =   limiting slenderness parameter for compact element

ℓ_a =   length of radius cut in beam flange for reduced beam section (RBS) connection design

ℓ_h =   distance from column centerline to centerline of hinge for RBS connection design

ℓ_n =   clear span measured face-to-face of supports

ℓ_u =   unsupported length of compression member

ℓ_w =   length of entire wall or of segment of wall considered in direction of shear force

μ =   coefficient of friction

Δ =   design story drift, shall be computed as the differences of the deflections at the center of mass at the top and bottom or the story under consideration. Note: Where ASD is used, Δ shall be computed using earthquake forces without dividing by 1.4, see §12.12

Δ =   design story drift

Δ_a =   allowable story drift, as obtained from Table 12.12-1 for any story

$\Delta_a A$ =   allowable story drift for structure A

$\Delta_a B$ =   allowable story drift for structure B

Ω_o =   system overstrength factor as given in Table 12.2-1

δ_x =   inelastic deflections of Level x – §12.8.6

δ_{AVE} =   the average of the displacements at the extreme points of the structure at Level x

δ_{MAX} =   the maximum displacement at Level x

δ_{XA} =   deflection at structure Level x of structure A

δ_{xe} =   the deflections determined by an elastic analysis of the seismic-force-resisting system

δ_M =   maximum of δ_x

δ_{M1}, δ_{M2} =   displacements of the adjacent building where δ_{M2} is at same level as δ_{M1}

δ_{YA} =   deflection at structure level y of structure A

δ_{YB} =   deflection at structure level y of structure B

θ =   stability coefficient – §12.8.7

Definitions

Active Fault/Active Fault Trace. A fault for which there is an average historic slip rate of 1 mm per year or more and geologic evidence of seismic activity within Holocene (past 11,000 years) times. Active fault traces are designated by the appropriate regulatory agency and/or registered design professional subject to identification by a geologic report.

Allowable Stress Design. A method of proportioning structural members, such that elastically computed stresses produced in the members by nominal loads do not exceed specified allowable stresses (also called working stress design).

Attachments, Seismic. Means by which components and their supports are secured or connected to the seismic-force-resisting system of the structure. Such attachments include anchor bolts, welded connections and mechanical fasteners.

Balcony, Exterior. An exterior floor projecting from and supported by a structure without additional independent supports.

Base. The level at which the horizontal seismic ground motions are considered to be imparted to the structure.

Base Shear. Total design lateral force or shear at the base.

Boundary Elements. Chords and collectors at diaphragm and shear wall edges, interior openings, discontinuities, and re-entrant corners.

Boundary Members. Portions along wall and diaphragm edges strengthened by longitudinal and transverse reinforcement and/or structural steel members.

Brittle. Systems, members, materials and connections that do not exhibit significant energy dissipation capacity in the inelastic range.

Cantilevered Column System. A structural system relying on column elements that cantilever from a fixed base and have minimal rotational resistance capacity at the top with lateral forces applied essentially at the top and are used for lateral resistance.

Collector. A diaphragm or shear wall element parallel to the applied load that collects and transfers shear forces to the vertical-force-resisting elements or distributes forces within a diaphragm or shear wall.

Component. A part or element of an architectural, electrical, mechanical, or structural system.
    Component, equipment. A mechanical or electrical component or element that is part of a mechanical and/or electrical system within or without a building system.
    Component, flexible. Component, including its attachments, having a fundamental period greater than 0.06 second.

Component, rigid. Component, including its attachments, having a fundamental period less than or equal to 0.06 second.

Confined Region. The portion of a reinforced concrete component in which the concrete is confined by closely spaced special transverse reinforcement restraining the concrete in directions perpendicular to the applied stress.

Coupling Beam. A beam that is used to connect adjacent concrete wall piers to make them act together as a unit to resist lateral forces.

Dead Loads. The weight of materials of construction incorporated into the building, including but not limited to walls, floors, roofs, ceilings, stairways, built-in partitions, finishes, cladding, and other similarly incorporated architectural and structural items, and fixed service equipment, including the weight of cranes.

Deck. An exterior floor supported on at least two opposing sides by an adjacent structure, and/or posts, piers, or other independent supports.

Deformability. The ratio of the ultimate deformation to the limit deformation.
> **High deformability element.** An element whose deformability is not less than 3.5 when subjected to four fully reversed cycles at the limit deformation.
> **Limited deformability element.** An element that is neither a low deformability nor a high deformability element.
> **Low deformability element.** An element whose deformability is 1.5 or less.

Deformation.
> **Limit deformation.** Two times the initial deformation that occurs at a load equal to 40 percent of the maximum strength.
> **Ultimate deformation.** The deformation at which failure occurs and which shall be deemed to occur if the sustainable load reduces to 80 percent or less of the maximum strength.

Design Earthquake. The earthquake effects that are 2/3 of MCE earthquake effects.

Design Strength. The product of the nominal strength and a resistance factor (or strength reduction factor).

Designated Seismic System. Those architectural, electrical, and mechanical systems and their components that require design in accordance with Chapter 13 that have a component importance factor, I_p, greater than 1.0.

Diaphragm, Flexible. A diaphragm is flexible for the purpose of distribution of story shear and torsional moment when the lateral deformation of the diaphragm is more than two times the average story drift of the associated story, determined by comparing the computed maximum in-plane deflection of the diaphragm itself under lateral force with the story drift of adjoining vertical lateral-force-resisting elements under equivalent tributary lateral force.

Diaphragm, Rigid. A diaphragm that does not conform to the definition of flexible diaphragm.

Displacement.

    Design Displacement. The design earthquake lateral displacement, excluding additional displacement due to actual and accidental torsion, required for design of the isolation system.

    Total Design Displacement. The design earthquake lateral displacement, including additional displacement due to actual and accidental torsion, required for design of the isolation system.

    Total Maximum Displacement. The maximum considered earthquake lateral displacement, including additional displacement due to actual and accidental torsion, required for verification of the stability of the isolation system or elements thereof, design of building separations, and vertical load testing of isolator unit prototype.

Displacement Restraint System. A collection of structural elements that limits lateral displacement of seismically isolated structures due to the maximum considered earthquake.

Duration of Load. The period of continuous application of a given load, or the aggregate of periods of intermittent applications of the same load.

Effective Damping. The value of equivalent viscous damping corresponding to energy dissipated during cyclic response of the isolation system.

Effective Stiffness. The value of the lateral force in the isolation system, or an element thereof, divided by the corresponding lateral displacement.

Element

    Ductile element. An element capable of sustaining large cyclic deformations beyond the attainment of its strength.

    Limited ductile element. An element that is capable of sustaining moderate cyclic deformations beyond the attainment of nominal strength without significant loss of strength.

    Nonductile element. An element having a mode of failure that results in an abrupt loss of resistance when the element is deformed beyond the deformation corresponding to the development of its nominal strength. Nonductile elements cannot reliably sustain significant deformation beyond that attained at their nominal strength.

Equipment Support. Those structural members or assemblies of members or manufactured elements, including braces, frames, lugs, snubbers, hangers, or saddles that transmit gravity load and operating load between the equipment and the structure.

Essential Facilities. Buildings and other structures that are intended to remain operational in the event of extreme environmental loading from flood, wind, snow, or earthquakes.

Factored Load. The product of a nominal load and a load factor.

Flexible Equipment Connections. Those connections between equipment components that permit rotational and/or translational movement without degradation of performance.

Frame.

 Braced frame. An essentially vertical truss, or its equivalent, of the concentric or eccentric type that is provided in a building frame system or dual frame system to resist shear.

 Concentrically braced frame (CBF). A braced frame in which the members are subjected primarily to axial forces.

 Eccentrically braced frame (EBF). A diagonally braced frame in which at least one end of each brace frames into a beam a short distance from a beam-column or from another diagonal brace.

 Ordinary concentrically braced frame (OCBF). A steel concentrically braced frame in which members and connections are designed for moderate ductility.

 Special concentrically braced frame (SCBF). A steel or composite steel and concrete concentrically braced frame in which members and connections are designed for ductile behavior.

Frame, Moment.

 Intermediate moment frame (IMF). A moment frame in which members and joints are capable of resisting forces by flexure as well as along the axis of the members.

 Ordinary moment frame (OMF). A moment frame in which members and joints are capable of resisting forces by flexure as well as along the axis of the members.

 Special moment frame (SMF). A moment frame in which members and joints are capable of resisting forces by flexure as well as along the axis of the members.

Frame System.

 Building frame system. A structural system with an essentially complete space frame system providing support for vertical loads. Seismic force resistance is provided by shear walls or braced frames.

 Dual frame system. A structural system with an essentially complete space frame system providing support for vertical loads. Seismic force resistance is provided by a moment-resisting frame and shear walls or braced frames.

 Space frame system. A structural system composed of interconnected members, other than bearing walls, that is capable of supporting vertical loads and that also may provide resistance to seismic forces.

Gravity Load (W). The total dead load and applicable portions of other loads as defined in §§12.7.2 and 12.14.8.1.

Hazardous Contents. A material that is highly toxic or potentially explosive and in sufficient quantity to pose a significant life-safety threat to the general public if an uncontrolled release were to occur.

Impact Load. The load resulting from moving machinery, elevators, craneways, vehicles, and other similar forces and kinetic loads, pressure, and possible surcharge from fixed or moving loads.

Importance Factor. A factor assigned to each structure according to its occupancy category as prescribed in §11.5.1.

Inverted Pendulum-type Structures. Structures that have a large portion of their mass concentrated near the top and, thus, have essentially one degree of freedom in horizontal translation. The structures are usually T-shaped with a single column supporting the beams or framing at the top.

Isolation Interface. The boundary between the upper portion of the structure, which is isolated, and the lower portion of the structure, which moves rigidly with the ground.

Isolation System. The collection of structural elements that includes individual isolator units, structural elements that transfer force between elements of the isolation system and connections to other structural elements.

Isolator Unit. A horizontally flexible and vertically stiff structural element of the isolation system that permits large lateral deformations under design seismic load. An isolator unit may be used either as part of or in addition to the weight-supporting system of the building.

Joint. A portion of a column bounded by the highest and lowest surfaces of the other members framing into it.

Limit State. A condition beyond which a structure or member becomes unfit for service and is judged to be no longer useful for its intended function (serviceability limit state) or to be unsafe (strength limit state).

Live Loads. Those loads produced by the use and occupancy of the building or other structure and do not include construction or environmental loads such as wind load, snow load, rain load, earthquake load, flood load, or dead load.

Live Loads (Roof). Those loads produced 1) during maintenance by workers, equipment, and materials; and 2) during the life of the structure by movable objects such as planters and by people.

Load and Resistance Factor Design (LRFD). A method of proportioning structural members and their connections using load and resistance factors such that no applicable limit state is reached when the structure is subjected to appropriate load combinations. The term "LRFD" is used in the design of steel and wood structures.

Load Factor. A factor that accounts for deviations of the actual load from the nominal load, for uncertainties in the analysis that transforms the load into a load effect, and for the probability that more than one extreme load will occur simultaneously.

Loads. Forces or other actions that result from the weight of building materials, occupants and their possessions, environmental effect, differential movement, and restrained dimensional changes. Permanent loads are those loads in which variations over time are rare or of small magnitude. Other loads are variable loads (see also "Nominal loads").

Loads Effects. Forces and deformations produced in structural members by the applied loads.

Maximum Considered Earthquake. The most severe earthquake effects considered by this code.

Nominal Loads. The magnitudes of the loads specified in this chapter (dead, live, soil, wind, snow, rain, flood, and earthquake.)

Nonbuilding Structure. A structure, other than a building, constructed of a type included in Chapter 15 and within the limits of §15.1.1.

Other Structures. Structures, other than buildings, for which loads are specified in this chapter.

***P*-delta Effect.** The second order effect on shears, axial forces and moments of frame members induced by axial loads on a laterally displaced building frame.

Panel (Part of a Structure). The section of a floor, wall, or roof located between the supporting frame of two adjacent rows of columns and girders or column bands of floor or roof construction.

Resistance Factor. A factor that accounts for deviations of the actual strength from the nominal strength and the manner and consequences of failure (also called strength reduction factor).

Seismic Design Category. A classification assigned to a structure based on its occupancy category and the severity of the design earthquake ground motion at the site, see §11.4.

Seismic-force-resisting system. The part of the structural system that has been considered in the design to provide the required resistance to the seismic forces prescribed herein.

Seismic Forces. The assumed forces prescribed herein, related to the response of the structure to earthquake motions, to be used in the design of the structure and its components.

Seismic Response Coefficient. Coefficient C_s, as determined from §12.8.

Shallow Anchors. Shallow anchors are those with embedment length-to-diameter ratios of less than 8.

Shear Panel. A floor, roof, or wall component sheathed to act as a shear wall or diaphragm.

Shear Wall. A wall designed to resist lateral forces parallel to the plane of the wall.

Shear Wall-frame Interactive System. A structural system that uses combinations of shear walls and frames designed to resist lateral forces in proportion to their rigidities, considering interaction between shear walls and frames on all levels.

Site Class. A classification assigned to a site based on the types of soils present and their engineering properties as defined in §11.4.2.

Site Coefficients. The values of F_a and F_v indicated in Tables 11.4-1 and 11.4-2, respectively.

Special Transverse Reinforcement. Reinforcement composed of spirals, closed stirrups, or hoops and supplementary cross-ties provided to restrain the concrete and qualify the portion of the component, where used, as a confined region.

Story Drift Ratio. The story drift divided by the story height.

Strength, Nominal. The capacity of a structure or member to resist the effects of loads, as determined by computations using specified material strengths and dimensions and formulas derived from accepted principles of structural mechanics or by field tests or laboratory tests of scaled models, allowing for modeling effects and differences between laboratory and field conditions.

Strength Design. A method of proportioning structural members such that the computed forces produced in the members by factored loads do not exceed the member design strength (also called load and resistance factor design.) The term "strength design" is used in the design of concrete and masonry structural elements.

Strength Required. Strength of a member, cross section, or connection required to resist factored loads or related internal moments and forces in such combinations as stipulated by these provisions.

Torsional Force Distribution. The distribution of horizontal seismic forces through a rigid diaphragm when the center of mass of the structure at the level under consideration does not coincide with the center of rigidity (sometimes referred to as a diaphragm rotation).

Toughness. The ability of a material to absorb energy without losing significant strength.

Wall, Load-bearing. Any wall meeting either of the following classifications:
1. Any metal or wood stud wall that supports more than 100 pounds per linear foot (1459 N/m) of vertical load in addition to its own weight.
2. Any masonry or concrete wall that supports more than 200 pounds per linear foot (2919 N/m) of vertical load in addition to its own weight.

Wall, Nonload-bearing. Any wall that is not a load-bearing wall.

Wind-restraint Seismic System. The collection of structural elements that provides restraint of the seismic-isolated-structure for wind loads. The wind-restraint system may be either an integral part of isolator units or a separate device.

Example i
Classification/Importance Factors   §11.5-1
Seismic Design Category   §11.6

Determine the importance factors and the seismic design category for a facility given the following information.

Type of occupancy – Elementary School with capacity greater than 250.
Per Table 1-1, Occupancy Category III

S_{DS} = 1.17
S_{D1} = 0.70
S_1 = 0.75

Determine the following.

1. **Building Occupancy Category and Importance Factors for Occupancy Category III and for a building to be used for an emergency shelter**

2. **Seismic Design Category (SDC)**

Calculations and Discussion   **Code Reference**

1. **Building category and importance factors.**

From Table 1-1, "Occupancy Category of Buildings and Other Structures for Flood, Wind, Snow, Earthquake and Ice Loads," the Occupancy Category for an Elementary School with an occupancy capacity greater than 250 is an Occupancy Category III. The Occupancy Category is used to determine the "Seismic Design Category," per Section 11.6-1. If the elementary school is to be used for an emergency shelter, the Occupancy Category is IV.

The importance factors for seismic loads are from Table 11.5-1. Importance factors for snow loads are from Table 7-4. Importance factors for wind loads are from Table 6-1.

Category	Seismic Factor I	Snow Factor I	Wind Factor I
III	1.25	1.1	1.15
IV	1.5	1.20	1.15

2. Seismic Design Category

All structures are assigned to a Seismic Design Category (SDC) based on their Occupancy Category and the spectral response acceleration coefficients S_{DS} and S_{D1}, irrespective of the fundamental period of vibration of the structure T. Each building and structure shall be assigned to the most severe SDC in accordance with Table 11.6-1 or 11.6-2 as follows.

Table 2.1 Occupancy Category vs Seismic Design Category

Nature of Occupancy	Occupancy Category	Table 11.6-1		Table 11.6-2		SDC USE*
		S_{DS}	SDC	S_{D1}	SDC	
School	III	1.17	D*	0.70	D*	E
Emergency Shelter	IV	1.17	D*	0.70	D*	F

Recall: $S_1 = 0.75\%$ for this table

*Note that for Occupancy Categories I, II, & III having S_1 equal to or greater than 0.75 (recall $S_1 = 0.75$), the building shall be assigned to SDC E. Also for Occupancy Category IV having $S_1 \geq 0.75$, the building shall be assigned to SDC F.

Example 1 ■ *Earthquake Load Combinations: Strength Design* §12.4.2.3

Example 1
Earthquake Load Combinations: Strength Design   §12.4.2.3

This example demonstrates the application of the strength design load combinations that involve the seismic load E given in §12.4.2.3. This will be done for the moment-resisting frame structure shown below.

$$S_{DS} = 1.10$$
$$I = 1.0$$
$$\rho = 1.3$$
$$f_1 = 0.5$$
Snow load $S = 0$

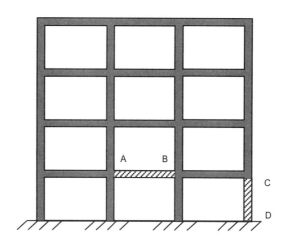

Beam A-B and Column C-D are elements of the special moment-resisting frame. Structural analysis has provided the following beam moments at A, and the column axial loads and moments at C due to dead load, office building live load, and left-to-right (→) and right-to-left (←) directions of lateral seismic loading.

	Dead Load D	Live Load L	Left-to-Right Seismic Load ($\rightarrow Q_E$)	Right-to-Left Seismic Load ($\leftarrow Q_E$)
Beam Moment at A	-100 kip-ft	-50 kip-ft	+120 kip-ft	-120 kip-ft
Column C-D Axial Load	+90 kips	+40 kips	+110 kips	-110 kips
Column Moment at C	+40 kip-ft	+20 kip-ft	+160 kip-ft	-160 kip-ft

Sign Convention: Positive moment induces flexural tension on the bottom side of a beam and at the right side of a column. Positive axial load induces compression. Note that for the particular location of Column C-D, the seismic Axial Load and Moment at C are both positive for the left-to-right (→) loading and are both negative for the right-to-left (←) loading. This is not necessarily true for the other elements of the structure.

Find the following.

1. Strength design seismic load combinations (Comb.)

2. Strength design moments at beam end A for seismic load combinations

3. Strength design interaction pairs of axial load and moment for the design of column section at C for seismic load combinations

Calculations and Discussion	*Code Reference*

1. Governing strength design seismic load combinations

$$1.2D + 1.0E + 0.5L \ldots \text{(Note } 0.2S = 0)$$ (Comb. 5)

$$0.9D + 1.0E$$ (Comb. 7)

where for a given type of load action such as moment M or axial load P

$$E = E_h + E_v$$ (Eq 12.4-1)

$$E_h = \rho Q_E$$ (Eq 12.4-3)

$$E_v = 0.2S_{DS}D$$ (Eq 12.4-4)

Combined, these yield

$$E = \rho Q_E + 0.2S_{DS}D$$ (Eq 12.4-3)

when the algebraic sign, ±, of Q_E is taken as the same as that for D, and

$$E = \rho Q_E - 0.2S_{DS}D$$

when the algebraic sign, ±, of Q_E is taken as opposite to that for D.

For the given values of : $\rho = 1.3$, $S_{DS} = 1.10$, the load combinations are

$$1.2D + 1.3Q_E + (0.2)(1.1)D + 0.5L = 1.42D + 1.3Q_E + 0.5L$$ (Comb. 5)

when the signs of Q_E and D are the same, and

$$1.2D + 1.3Q_E - (0.2)(1.1)D + 0.5L = 0.98D + 1.3Q_E + 0.5L$$ (Comb. 5)

when the signs of Q_E and D are opposite.

$$0.9D + 1.3Q_E + (0.2)(1.1)D = 1.12D + 1.3Q_E$$ (Comb. 7)

when the signs of Q_E and D are the same, and

Example 1 ■ *Earthquake Load Combinations: Strength Design*   §12.4.2.3

$$0.9D + 1.3Q_E - (0.2)(1.1)D = 0.68D + 1.3Q_E \qquad \text{(Comb. 7)}$$

when the signs of Q_E and D are opposite.

By inspection, the governing seismic load combinations are

$$1.42D + 1.3Q_E + 0.5L$$

when the signs of Q_E and D are the same,

$$0.68D + 1.3Q_E$$

when the signs of Q_E and D are opposite.

---

2. **Strength design moments at beam end A for seismic load combinations**

a. For the governing load combination when the signs of Q_E and D are the same

$$1.42D + 1.3Q_E + 0.5L$$

with $D = M_D = -100$, $Q_E = M_{QE} = -120$, and $L = M_L = -50$

$$M_A = 1.42\,(-100) + 1.3\,(-120) + 0.5(-50) = -323 \text{ kip-ft}$$

b. For the governing load combination when the signs of Q_E and D are opposite

$$0.68D + 1.3Q_E$$

with $D = M_D = -100$ and $Q_E = 120$

$$M_A = 0.68(-100) + 1.3(120) = 88 \text{ kip-ft}$$

∴ Beam section at A must be designed for

$$M_A = -323 \text{ kip-ft and } + 88 \text{ kip-ft}$$

3. **Strength design interaction pairs of axial load and moment for the design of column section at C for seismic load combinations**

The seismic load combinations using the definitions of E given by Equations 12.4-1 through 12.4-4 can be used for the design requirement of a single action such as the moment at beam end A, but they cannot be used for interactive pairs of actions such as the axial load and moment at the column section C. These pairs must occur simultaneously because of a common load combination. For example, both the axial load and the moment must be due to a common direction of the lateral seismic loading and a common sense of the vertical seismic acceleration effect represented by $0.2\,S_{DS}D$. There can be cases where the axial load algebraic signs are the same for Q_E and D, while the moment algebraic signs are different. This condition would prohibit the use of the same load combination for both axial load and moment.

To include the algebraic signs of the individual actions, the directional property of the lateral seismic load effect Q_E, and the independent reversible property of the vertical seismic load effect $0.2\,S_{DS}D$, it is proposed to use

$$E = \rho(\rightarrow Q_E) \pm 0.2\,S_{DS}D, \text{ and } \rho(\leftarrow Q_E) \pm 0.2\,S_{DS}D.$$

The resulting set of combinations is

$$1.2D + \rho(\rightarrow Q_E) + 0.2\,S_{DS}D + L$$

$$1.2D + \rho(\rightarrow Q_E) - 0.2\,S_{DS}D + L$$

$$1.2D + \rho(\leftarrow Q_E) + 0.2\,S_{DS}D + L$$

$$1.2D + \rho(\leftarrow Q_E) - 0.2\,S_{DS}D + L$$

$$0.9D + \rho(\rightarrow Q_E) + 0.2\,S_{DS}D$$

$$0.9D + \rho(\rightarrow Q_E) - 0.2\,S_{DS}D$$

$$0.9D + \rho(\leftarrow Q_E) + 0.2\,S_{DS}D$$

$$0.9D + \rho(\leftarrow Q_E) - 0.2\,S_{DS}D$$

(Note: a factor of 0.5 applies to L if $L \le 100$ psf [except at garages and public assembly areas])

For the specific values of $\rho = 1.3$ and $S_{DS} = 1.10$, the load combinations provide the following values for M_A, and the interaction pair P_C and M_C. Note that the interaction pair P_C and M_C must occur simultaneously at a specific load combination of gravity load, and lateral and vertical seismic load effects. The interaction design of the column section must satisfy all of the eight pairs of P_C and M_C from the seismic load

Example 1 ■ *Earthquake Load Combinations: Strength Design* §*12.4.2.3*

combinations along with the pairs from the gravity load combinations and wind load combinations.

Combination	M_A kip-ft	P_C kips	and	M_C kip-ft
$1.42D + 1.3\,(Q_E) + 0.5L$	-35	+268.8	and	+242.8
$0.98D + 1.3\,(Q_E) + 0.5L$	-9	+26.8	and	+225.2
$1.42D - 1.3\,(Q_E) + 0.5L$	-299	+229.2	and	-109.2
$0.98D - 1.3\,(Q_E) + 0.5L$	-255	-12.8	and	-126.8
$1.12D + 1.3\,(Q_E)$	+20	+221.8	and	+220.8
$0.68D + 1.3\,(Q_E)$	+64	+182.2	and	+203.2
$1.12D - 1.3\,(Q_E)$	-244	-20.2	and	-131.2
$0.68D - 1.3\,(Q_E)$	-200	-59.8	and	-148.8

The governing values are underlined for M_A [same as determined in Part (2)] and for the interaction pairs of P_C and M_C required for the design of the column section at C.

Commentary

The eight seismic load combinations resulting from the proposed definition of E provide an automatic method of considering the individual algebraic signs of the load actions, the direction of the lateral seismic load, and the independent ± action of $0.2\,S_{DC}D$. There is no need to use the "same sign" and "opposite sign" limitations of Equations 12.4-2 and 12.4-3 since all possible combinations are represented. This is important for interactive pairs of actions that must be evaluated for a common load combination.

When the Modal Response Spectrum Analysis procedure of §12.9 is used, the algebraic signs of seismic load actions are lost because of the process of combining the individual modal responses. The signs to be used for an interaction pair of actions due to a given direction of lateral loading can be obtained from the primary mode response where the primary mode is the mode having the largest participation factor for the given direction of lateral seismic loading. Or, alternatively, the signs can be obtained from the equivalent lateral force procedure of §12.8.

Example 2
Combinations of Loads   §2.4

The code permits the use of allowable stress design for the design of wood members and their fastenings (ASCE/SEI 7-05 §2.4 and §12.4.2.3). Section 2.4 defines the basic load combinations for allowable stress design.

This example illustrates the application of this method for the plywood shear wall shown below. The wall is a bearing and shear wall in a light wood framed building.

The following information is given.

Seismic Design Category B

$$I = 1.0$$
$$\rho = 1.0$$
$$S_{DS} = 0.3$$

$$E = E_h = \delta Q_E = 4 \text{ kips (seismic}$$
force due to the base shear
determined from §12.4.2)

Gravity loads
Dead $w_D = 0.3$ klf (tributary dead
load, including weight of
wall
Live $w_L =$ (roof load supported by
other elements)

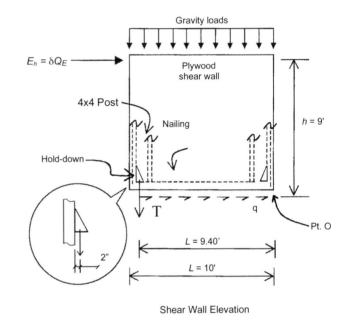

Shear Wall Elevation

Moment arm from center of post to center of hold-down bolt
$$L = 10 \text{ ft} - (3.5 + 2.0 + 3.5/2) = 10 \text{ ft} - 7.25 \text{ in} = 9.4 \text{ ft}$$

Determine the required design loads for shear capacity q and hold-down capacity T for the following load combinations.

1. Basic allowable stress design

Example 2 ■ *Combinations of Loads* §2.4

Calculations and Discussion	Code Reference

1. Basic allowable stress design §12.4.2.3

The governing load combinations for basic allowable stress design are Basic ASD Combinations 5, 6, and 8, as modified in §12.4.2.3. These are used without the usual one-third stress increase.

§12.4.2 defines the seismic load effect E for use in load combinations as

$$E = E_h + E_v \qquad \text{(Eq 12.4-1)}$$

$$= \delta Q_E + 0.2 \, S_{DS}D \qquad \text{(Eq 12.4-3)}$$

$$= Q_E + 0.06D \text{ when } D \text{ and } Q_E \text{ are in the same sense} \qquad \text{(Eq 12.4-4)}$$

and $\quad E = \rho Q_E - 0.2 S_{DS}D$

$$= Q_E - 0.06D \text{ when } D \text{ and } Q_E \text{ have opposite sense}$$

For ASD Basic Combination 5 the load combination is:

$$D + 0.7E \qquad \text{(Comb. 5)}$$

$$= D(1.0) + 0.7 \, (0.6D + Q_E)$$

$$= (\overset{1.42}{\cancel{1.042}})D + 0.7Q_E \text{ for } D \text{ and } Q_E \text{ with the same sense}$$

and $\quad D(1.0) + 0.7 \, (-0.6D - Q_E)$

$$= 0.958D - 0.7Q_E \text{ for } D \text{ and } Q_E \text{ with opposite sense}$$

For ASD Basic Combination 6 the load combination is:

$$D + 0.75(0.7E) + 0.75 \, (L + L_r) \qquad \text{(Comb. 6)}$$

$$= D(1.0 + (0.75)(0.7)(0.06)) + (0.75)(0.70)(1.0)Q_E + 0.75 \, L_r$$

$$= 1.032D + 0.75L_r + 0.525 \, Q_E \text{ for } D \text{ and } Q_E \text{ with the same sense}$$

$$= -0.968D + 0.75 \, L_r - 0.525 \, Q_E \text{ for } D \text{ and } Q_E \text{ with the opposite sense}$$

For ASD Basic Combination 8 the load combination is:

$$0.6D + 0.7E \qquad \text{(Comb. 8)}$$

$$= D(0.06) + 0.7(1.0) \ Q_E + 0.7(0.06)D$$

$$= (0.6 + 0.042)D + 0.7Q_E$$

$$= 0.642D + 0.7Q_E \text{ for } D \text{ and } Q_E \text{ in the same sense}$$

$$= (0.6 - 0.042)D - 0.7Q_E$$

$$= 0.558D - 0.7Q_E \text{ for } D \text{ and } Q_E \text{ in the opposite sense}$$

For the determination of design shear capacity, dead load and live load are not involved, and all load combinations reduce to

$$0.7Q_E$$

For the design hold-down tension capacity the governing load combination is

$$0.558D - 0.7Q_E$$

For the wall boundary element compression capacity, the governing load combination would be

$$1.042D + 0.7Q_E$$

a. Required unit shear capacity *q*

Base shear and the resulting element seismic forces Q_E determined under §12.8.1 are on a strength design basis. For allowable stress design, Q_E must be factored by 0.7 as indicated.

For design shear capacity the seismic load effect is

$$Q_E = 4000 \text{ lb}$$

For the governing load combination of $0.7Q_E$, the design unit shear is

$$q = \frac{0.7Q_E}{L} = \frac{0.7(4000)}{10 \text{ ft}} = 280 \text{ plf}$$

This unit shear is used to determine the plywood thickness and nailing requirements from IBC Table 2306.4.1, which gives allowable shear values for short-time duration loads due to wind or earthquake. For example, select 15/32 structural I sheeting (plywood) with 10d common nails having a minimum penetration of 1-1/2 inches

Example 2 ■ *Combinations of Loads*   §2.4

into 2x members with 6-inch spacing of fasteners at panel edges; allowable shear of 340 plf.

| **b.** | **Required hold-down tensile capacity *T*** |

Taking moments about point O at center of post at right side of wall with $E_h = \delta Q_E = 4000$ lb, the value of the hold-down tension force *T* due to horizontal seismic forces is computed

$$0.558(300 \text{ plf})10 \text{ ft}(5 \text{ ft} - \frac{3.5}{2(12)}) - 0.7(4000 \text{ lb})(9 \text{ ft}) + T(9.4 \text{ ft}) = 0$$

Thus:

$$8125.88 \text{ lb ft} - 25{,}200 \text{ lb ft} + 9.4 \text{ ft}(T) = 0$$

$$T = 1816.39 \text{ lb tension}$$

Similarly the boundary element compression capacity is computed

$$1.042(300 \text{ plf})10 \text{ ft} (5 \text{ ft} - \frac{3.5}{2(12)}) + 0.7 (4000 \text{ lb})(9 \text{ ft}) - C(9.4 \text{ ft}) = 0$$

Thus:

$$15{,}174 \text{ lb ft} + 25{,}200 \text{ lb ft} - 9.4 \text{ ft } C = 0$$

$$C = 4295 \text{ lb compression}$$

The tension value is used for the selection of the pre-manufactured hold-down elements. Manufacturer's catalogs commonly list hold-down sizes with their "1.33 × allowable" capacity values. Here the 1.33 value represents the allowed Load Duration factor for resisting seismic loads. This is not considered a stress increase (although it has the same effect). Therefore, the catalog "1.33 × allowable" capacity values may be used to select the appropriate hold-down element.

Commentary

Equations 12.4-1 and 12.4-2 for *E* create algebraic sign problems in the load combinations. It would be preferable to use

$$E = \rho Q_E + 0.2 S_{DS}D$$

and use $\pm E$ in the load combinations.

Example 3
Design Spectral Response Accelerations   §11.4

For a given building site, the <u>maximum considered earthquake</u> spectral response accelerations S_s at short periods, and S_1 at 1-second period are given by the acceleration contour maps in §22. This example illustrates the general procedure for determining the design spectral response parameters S_{DS} and S_{D1} from the mapped values of S_S and S_1. The parameters S_{DS} and S_{D1} are used to calculate the design base shear in §12.8 and the Design Response Spectrum in §11.4.5.

Note that by far the most accurate, easiest, and most efficient way to obtain the spectral design values is to use the USGS website (*www.eqhazmaps.usgs.gov*). Given the longitude and latitude of the site, the website provides values of S_S and S_1. The site longitude and latitude can be obtained from an internet site such as "*www.geocode.com*" by simply inputting the address.

From "*www.geocode.com*" it is determined that a building site near Sacramento, California is located at Latitude 38.123° North and Longitude –121.123° (or 121.123° west). The soil profile is Site Class D.

Determine the following.

1. Maximum considered earthquake spectral response accelerations S_s and S_1

2. Site coefficients and adjusted maximum considered earthquake spectral response acceleration parameters S_{MS} and S_{M1}

3. Design spectral response acceleration parameters S_{DS} and S_{D1}

4. Plot the general procedure response spectrum

5. Calculation of seismic response coefficient C_s
Given: soil site class D, $R = 6$, $T = 0.60$ sec, and $I = 1.0$

| *Calculations and Discussion* | *Code Reference* |

1. Maximum considered earthquake spectral response accelerations §11.4.1

For the given position (Near Sonora – NW of Sacramento, California) of 38° North (Latitude = 38.123°) and 121.123° West (Longitude = – 121.123°), USGS provides the values of

$$S_S = 57.3\%g = 0.573g$$

$$S_1 = 23.0\%g = 0.230g$$

2. Site coefficients and adjusted maximum considered earthquake spectral response accelerations §11.4.3

From the USGS for the given site class *D*, and $S_S = 0.573g$, $S_1 = 0.230g$, the site coefficients are as follows

$$F_a = 1.34 \hspace{4cm} \text{T 11.4-1}$$

$$F_v = 1.94 \hspace{4cm} \text{T 11.4-2}$$

The adjusted maximum considered earthquake spectral response accelerations (based on §11.4.3) are also given on the CD ROM as follows

$$S_{MS} = F_a S_s = 1.34(0.573g) = 0.768g \hspace{2cm} \text{(Eq 11.4-1)}$$

$$S_{M1} = F_v S_1 = 1.94(0.230g) = 0.446g \hspace{2cm} \text{(Eq 11.4-2)}$$

| **3.** | **Design spectral response acceleration parameters** | **§11.4.4** |

$$S_{DS} = \frac{2}{3} S_{MS} = \frac{2}{3} (0.768g) = 0.512g \qquad \text{(Eq 11.4-3)}$$

$$S_{D1} = \frac{2}{3} S_{M1} = \frac{2}{3} (0.446g) = 0.297g \qquad \text{(Eq 11.4-4)}$$

| **4.** | **General procedure response spectrum** | **§11.4.5** |

For periods less than or equal to T_o, the design spectral response shall be given by

$$S_a = 0.6 \frac{S_{DS}}{T_o} T + 0.4 S_{DS} \qquad \text{(Eq 11.4-5)}$$

For periods greater than or equal to T_o and less than or equal to T_s, the design spectral response acceleration S_a shall be taken equal to S_{DS}

For periods greater than T_s, and less than T_L, the design spectral response acceleration S_a shall be given by

$$S_a = (S_{D1}) / T \qquad \text{(Eq 11.4-6)}$$

Where: $T_o = 0.20 (S_{D1} / S_{DS})$

$\qquad = 0.2 (0.297 / 0.512)$

$\qquad = 0.12$ sec

$T_s = S_{D1} / S_{DS}$

$\qquad = 0.297 / 0.512$

$\qquad = 0.58$ sec

$T_L = 8$ sec $\qquad \text{(F 22-15)}$

Thus:

T = Period	S_a/g	Computation for S_a
0.00	0.20	0.4 (0.51)
0.12	0.51	0.51
0.58	0.51	0.51
0.80	0.41	0.30 / 0.8
1.00	0.25	0.30 / 1.2
1.20	0.21	0.30 / 1.4
1.40	0.19	0.30 / 1.6
1.60	0.17	0.30 / 1.8
2.00	0.15	0.30 / 2.0

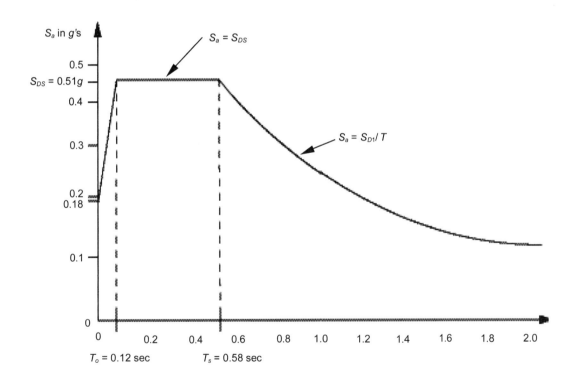

General Procedure Response Spectrum

5. Calculation of seismic response coefficient C_s (Recall Soil Site Class D, $I = 1.0$ and $T = 0.60$)   §12.8.1

The seismic response coefficient shall be determined by

$$C_s = S_{DS} / (R/I)$$ (Eq 12.8-2)

$$= 0.546 / (6.0/1.0)$$

$$= 0.091 \ldots \text{Governs}$$

The value of C_s need not exceed

$$C_s = S_{D1} / (R/I) \, T$$ (Eq 12.8-3)

$$= 0.30 / (6.0/1.0) \, (0.6)$$

$$= 0.085$$

for $T \leq T_L$

But shall not be taken less than

$$C_s = 0.01$$ (Eq 12.8-5)

where $S_1 \geq 0.6g$ C_s shall not be less than

$$C_s = 0.5S_1 / (R/I)$$ (Eq 12.8-6)

Introduction to
Vertical Irregularities   §12.3.2.2

Table 12.3-2 defines vertical structural irregularities and assigns analysis and design procedures to each type and seismic design category. These irregularities can be divided into two categories. The first, dynamic force-distribution irregularities, which are Types 1a, 1b, 2, and 3. The second, irregularities in load path or force transfer, which are Types 4 and 5. The vertical irregularities are

1a. Stiffness Soft Story Irregularity

1b. Stiffness Extreme Soft Story Irregularity

2. Weight (mass) irregularity

3. Vertical geometric irregularity

4. In-plane discontinuity in vertical lateral-force-resisting element

5a. Discontinuity in Lateral Stength – Weak Story Irregularity

5b. Discontinuity in Lateral Strength – Extreme Weak Story Irregularity

Structures in Seismic Design Categories D, E, and F possessing dynamic force distribution irregularities shall be analyzed using the dynamic analysis procedure (or modal analysis procedure) prescribed in §12.7. (Refer to Table 12.6.1) Structure Description 3. The vertical force distribution provided by §12.8.3 may be assumed to be adequate for structures lacking vertical irregularity Types 1a, 1b, 2, and 3. However, stiffness and mass discontinuities may significantly affect the vertical distribution of forces and, for this reason the modal analysis procedure, which can account for these discontinuities, is necessary.

Although designers may opt to use the dynamic analysis procedure and bypass checks for irregularity Types 1a, 1b, 2, and 3, the reference sections listed in Table 12.3-2 should still be checked for limitations and design requirements. Note that §12.3.3.1 prohibits structures with vertical irregularity Types 1b, 5a, or 5b for Seismic Design Categories E and F.

Regular structures are assumed to have a reasonably uniform distribution of inelastic behavior in elements throughout the lateral-force-resisting system. When vertical irregularity Types 4 and 5 exist, there is the possibility of having localized concentrations of excessive inelastic deformations due to the irregular load path or weak story. In this case, the code prescribes additional strengthening to correct the deficiencies for structures in certain seismic design categories (SDCs). In the case of vertical irregularity Type 5b, limits are placed on the building height for all SDCs except SDC A.

Example 4
Vertical Irregularity Type 1a and Type 1b   §12.3.2.2

A Seismic Design Category D five-story concrete special moment-resisting frame is shown below. The code-prescribed lateral forces F_x from Equation 12.8-11 have been applied and the corresponding floor level displacements δ_{xe} at the floors' centers-of-mass have been determined as shown below.

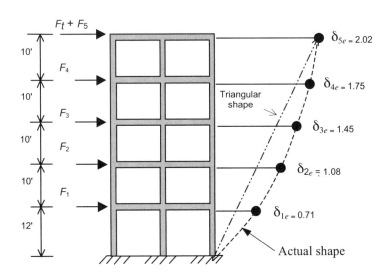

---

1. Determine if a Type 1a vertical irregularity from Table 12.3-2 (Stiffness-Soft Story Irregularity) exists in the first story

---

Calculations and Discussion   **Code Reference**

---

1. To determine if this is a Type 1a vertical irregularity (Stiffness-Soft Story Irregularity) there are two tests

1. The lateral story stiffness is less than 70 percent of that of the story above.

2. The lateral story stiffness is less than 80 percent of the average stiffness of the three stories above.

Example 4 ■ *Vertical Irregularity Type 1* §*12.3.2.2*

If the stiffness of the story meets at least one of the two criteria above, the structure is deemed to have a soft story, and a modal analysis (§12.9) is generally required by Table 12.6-1.

The definition of soft story in the code compares values of the lateral stiffness of individual stories. Generally, it is not practical to use stiffness properties unless these can be easily determined. There are many structural configurations where the evaluation of story stiffness is complex and is often not an available output from computer programs. Recognizing that the basic intent of this irregularity check is to determine if the lateral-force distribution will differ significantly from the pattern prescribed by §12.8.3, which assumes a prescribed shape for the first dynamic mode of response, this type of irregularity can also be determined by comparing values of drift ratios due to the prescribed lateral forces. This deformation comparison may even be more effective than the stiffness comparison because the shape of the first mode shape is often closely approximated by the structure displacements due to the specified §12.8.3 force pattern. Floor level displacements and corresponding story-drift ratios are directly available from computer programs. To compare displacements rather than stiffness, it is necessary to use the reciprocal of the limiting percentage ratios of 70 and 80 percent as they apply to story stiffness, or reverse their applicability to the story or stories above. The following example shows this equivalent use of the displacement properties.

From the given displacements, story drifts and the story-drift ratio's values are determined. The story-drift ratio is the story drift divided by the story height. These story-drift ratios will be used for the required comparisons because they better represent the changes in the slope of the mode shape when there are significant differences in interstory heights. (Note: story displacements can be used if the story heights are nearly equal.)

In terms of the calculated story-drift ratios, the soft story occurs when one of the following conditions exists.

When 70 percent of $\dfrac{\delta_{1e}}{h_1}$ exceeds $\dfrac{\delta_{2e}-\delta_{1e}}{h_2}$

    or

When 80 percent of $\dfrac{\delta_{1e}}{h_1}$ exceeds $\dfrac{1}{3}\left[\dfrac{\left(\delta_{2e}-\delta_{1e}\right)}{h_2}+\dfrac{\left(\delta_{3e}-\delta_{2e}\right)}{h_3}+\dfrac{\left(\delta_{4e}-\delta_{3e}\right)}{h_4}\right]$

the story-drift ratios are determined as

$$\frac{\Delta_1}{h_1}=\frac{\delta_{1e}}{h_1}=\frac{(0.71-0)}{144}=0.00493$$

$$\frac{\Delta_2}{h_2}=\frac{\delta_{2e}-\delta_{1e}}{h_2}=\frac{(1.08-0.71)}{120}=0.00308$$

$$\frac{\Delta_3}{h_3}=\frac{\delta_{3e}-\delta_{2e}}{h_3}=\frac{(1.45-1.08)}{120}=0.00308$$

$$\frac{\Delta_4}{h_4} = \frac{\delta_{4e} - \delta_{3e}}{h_4} = \frac{(1.75 - 1.45)}{120} = 0.00250$$

$$\frac{1}{3}(0.00308 + 0.00308 + 0.00250) = 0.00289$$

Checking the 70-percent requirement:

$$0.70\left(\frac{\delta_{1e}}{h_1}\right) = 0.70(0.00493) = 0.00345 > 0.00308 \dots \text{NG}$$

∴ Soft story exists. . .

Note that 70 percent of first story drift is larger than second story drift. Alternately: $0.00493 > (0.00308 \times 1.30 = 0.0040) \dots$ thus soft story.
Also note that structural irregularities of Types 1a, 1b, or 2 in Table 12.3-2 do not apply where no story-drift ratio under design lateral force is greater than 130 percent of the story-drift ratio of the next story above, §12.3.2.2, Exception 1.

Checking the 80-percent requirement:

$$0.80\left(\frac{\delta_{1e}}{h_1}\right) = 0.80(0.00493) = 0.00394 > 0.00289 \dots \text{NG}$$

∴ Soft story exists. . . condition 1a

Alternately: $0.00493 > (0.00289 \times 1.20 = 0.00347) \dots$ thus soft story.

Check for extreme soft story, (Vertical Structural Irregularity, Type 1b)

Checking the 60-percent requirement:

$0.60(0.00493) = 0.002958 < 0.00308 \dots o.k.$

Alternately: $0.00493 > (0.00308 \times 1.4 = 0.004312) \dots o.k.$

Checking the 70-percent requirement:

$0.70\,(0.00493) = 0.003451 > 0.00289 \dots \text{NG}$

Alternately: $0.00493 > (0.00289 \times 1.3 = 0.00375) \dots \text{NG}$

Thus: Stiffness-Extreme Soft Story exists – condition 1b.

Example 4 ■ *Vertical Irregularity Type 1* §12.3.2.2

Recall from Table 12.3-2 for 1b, extreme soft story, reference §12.3.3.1. This building is SDC D, and is permitted. Structures having SDC E or F and also having vertical irregularity Type 1b <u>shall not</u> be permitted.

Commentary

Section 12.8.6 requires that story drifts be computed using the maximum inelastic response displacements δ_x, which include the deflection amplification factor C_d

$$\delta_x = \frac{C_d \delta_{xe}}{I}$$

(Eq 12.8-15)

However, for the purpose of the story drift, or story-drift ratio, comparisons needed for soft-story determination, the displacement δ_{xe} due to the design seismic forces can be used as in this example. In the example above, only the first story was checked for possible soft-story vertical irregularity. In practice, all stories must be checked, unless a modal analysis is performed. It is often convenient to create tables to facilitate this exercise, see Tables 4.1 and 4.2.

Table 4.1 Soft-Story Status 1a

Level	Story Displacement	Story Drift	Story-drift Ratio	0.8x (Story-drift Ratio)	0.7x (Story-drift Ratio)	Avg. of Story-drift Ratio of Next 3 Stories	Soft Story Status 1a
5	2.02 in	0.27 in	0.00225	0.00180	0.00158	—	No
4	1.75	0.30	0.00250	0.00200	0.00175	—	No
3	1.45	0.37	0.00308	0.00246	0.00216	—	No
2	1.08	0.37	0.00308	0.00246	0.00216	0.00261	No
1	0.71	0.71	0.00493	0.00394	0.00345	0.00289	Yes

Table 4.2 Soft-Story Status 1b

Level	Story Displacement	Story Drift	Story-drift ratio	0.7x (Story-drift Ratio)	0.6x (Story-drift Ratio)	Avg. of Story-drift Ratio of Next 3 Stories	Soft Story Status1b
5	2.02 in	0.27 in	0.00225	0.00158	0.00135	—	No
4	1.75	0.30	0.00250	0.00175	0.00150	—	No
3	1.45	0.37	0.00308	0.00216	0.00185	—	No
2	1.08	0.37	0.00308	0.00216	0.00185	0.00261	No
1	0.71	0.71	0.00493	0.00345	0.00296	0.00289	Yes

Example 5
Vertical Irregularity Type 2   §12.3.2.2

The five-story special moment frame office building has a heavy utility equipment installation at Level 2. This results in the floor weight distribution shown below.

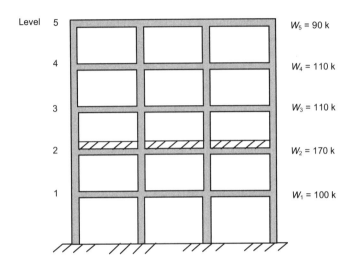

Level	
5	$W_5 = 90$ k
4	$W_4 = 110$ k
3	$W_3 = 110$ k
2	$W_2 = 170$ k
1	$W_1 = 100$ k

---

1. **Determine if there is a Type 2 vertical weight (mass) irregularity**

Calculations and Discussion   **Code Reference**

A weight, or mass, vertical irregularity is considered to exist when the effective mass of any story is more than 150 percent of the effective mass of an adjacent story. However, this requirement does not apply to the roof if the roof is lighter than the floor below. Note that it does apply if the roof is heavier than the floor below.

Checking the effective mass of Level 2 against the effective mass of Levels 1 and 3

At Level 1

$$1.5 \times W_1 = 1.5(100 \text{ kips}) = 150 \text{ kips}$$

At Level 3

$$1.5 \times W_3 = 1.5(110 \text{ kips}) = 165 \text{ kips}$$

$$W_2 = 170 \text{ kips} > 150 \text{ kips}$$

∴ Weight irregularity exists.

Example 5 ■ *Vertical Irregularity Type 2*   *§12.3.2.2*

Commentary

As in the case of vertical irregularity Type 1a or 1b, this Type 2 irregularity also results in a primary mode shape that can be substantially different from the triangular shape and lateral load distribution given by §12.8.3. Consequently, the appropriate load distribution must be determined by the modal analysis procedure of §12.9, unless the irregular structure is not more than two stories and is Occupancy Category I or II (see Table 12.6-1).

Example 6
Vertical Irregularity Type 3   §12.3.2.2

The lateral-force-resisting system of the five-story special moment frame building shown below has a 25-foot setback at the third, fourth, and fifth stories.

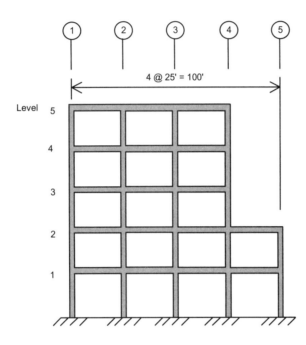

---

1. **Determine if a Type 3 vertical irregularity (vertical geometric irregularity) exists**

Calculations and Discussion   **Code Reference**

A vertical geometric irregularity is considered to exist where the horizontal dimension of the lateral-force-resisting system in any story is more than 130 percent of that in the adjacent story. One-story penthouses are not subject to this requirement.

In this example, the setback of Level 3 must be checked. The ratios of the two levels are

$$\frac{\text{Width of Level 2}}{\text{Width of Level 3}} = \frac{(100 \text{ ft})}{(75 \text{ ft})} = 1.33$$

133 percent $>$ 130 percent

∴ Vertical geometric irregularity exists.

Example 6 ■ *Vertical Irregularity Type 3*   *§12.3.2.2*

Commentary

The more than 130-percent change in width of the lateral-force-resisting system between adjacent stories could result in a primary mode shape that is substantially different from the shape assumed for proper applications of Equation 12.8-11. If the change is a decrease in width of the upper adjacent story (the usual situation), the mode shape difference can be mitigated by designing for an increased stiffness in the story with a reduced width.

Similarly, if the width decrease is in the lower adjacent story (the unusual situation), the Type 1a soft-story irregularity can be avoided by a proportional increase in the stiffness of the lower story. However, when the width decrease is in the lower story, there could be an overturning moment-load-transfer discontinuity that would require a dynamic analysis per Table 12.6-1.

Note that if the frame elements in the bay between lines 4 and 5 were not included as part of the designated lateral-force-resisting system, the vertical geometric irregularity would not exist.

Example 7
Vertical Irregularity Type 4   §12.3.2.2

A concrete building has the building frame system shown below. The shear wall
between lines A and B has an in-plane offset from the shear wall between lines C
and D.

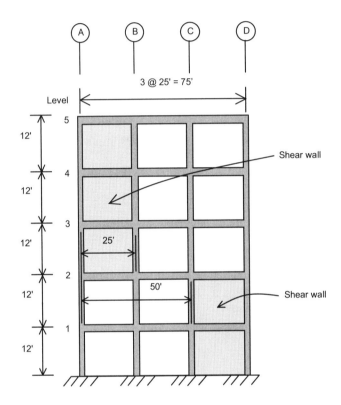

1. | **Determine if there is a Type 4 vertical irregularity (in-plane discontinuity)
in the vertical lateral-force-resisting element**

Calculations and Discussion   *Code Reference*

A Type 4 vertical irregularity exists when there is an in-plane offset of the lateral-force-
resisting elements greater than the length of those elements. In this example, the left
side of the upper shear wall (between lines A and B) is offset 50 feet from the left
side of the lower shear wall (between lines C and D). This 50-foot offset is greater
than the 25-foot length of the offset wall elements.

∴ In-plane discontinuity exists.

Example 7 ▪ *Vertical Irregularity Type 4* §*12.3.2.2*

Commentary

The intent of this irregularity check is to provide correction of force transfer or load-path deficiencies. It should be noted that any in-plane offset, even those less than or equal to the length or bay width of the resisting element, can result in an overturning moment-load-transfer discontinuity that requires the application of §12.3.3.3. When the offset exceeds the length of the resisting element, there is also a shear transfer discontinuity that requires application of §12.3.3.4 for the strength of collector elements along the offset. In this example, the columns under wall A-B are subject to the provisions of §12.3.3.3, and the collector element between lines B and C at Level 2 is subject to the provisions of §12.3.3.4.

Example 8
Vertical Irregularity Type 5a   §12.3.2.2

A concrete bearing-wall building has the typical transverse shear-wall configuration shown below. All walls in this direction are identical, and the individual piers have the shear contribution given below. Then, V_n is the nominal shear strength calculated in accordance with Chapter 19, and V_m is defined herein as the shear corresponding to the development of the "nominal flexure strength also calculated in accordance with Chapter 19." Note that V_m is not defined in ACI or Chapter 19.

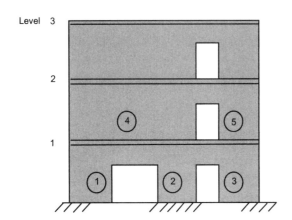

PIER	V_n	V_m
1	20 kips	30 kips
2	30	40
3	15	10
4	80	120
5	15	10

1. **Determine if a Type 5 vertical irregularity (discontinuity in capacity- weak-story) condition exists**

Calculations and Discussion   **Code Reference**

A Type 5a weak-story discontinuity in capacity exists when the story strength is less than 80 percent of that in the story above. The story strength is the total strength of all seismic-force-resisting elements sharing the story shear for the direction under consideration.

Using the smaller values of V_n and V_m given for each pier, the story strengths are

First story strength = 20 + 30 + 10 = 60 kips

Second story strength = 80 + 10 = 90 kips

Check if first-story strength is less than 80 percent of that of the second story.

60 kips < 0.8(90) = 72 kips

∴ Weak story condition exists.

Example 8 ■ *Vertical Irregularity Type 5a*   *§12.3.2.2*

Check if first-story strength is less than 65 percent of that of the second story (Irregularity Type 5b).

$$60 \text{ kips} < 0.65(90 \text{ kips}) = 58.5 \text{ kips}$$
$$\therefore \; 60 \text{ kips} > 58.5 \text{ kips}$$

∴   Therefore the lower story is not an extreme soft story, Irregularity Type 5b.

Commentary

This irregularity check is to detect any concentration of inelastic behavior in one supporting story that can lead to the loss of vertical load capacity. Elements subject to this check are the shear-wall piers (where the shear contribution is the lower of either the shear at development of the flexural strength, or the shear strength), bracing members and their connections, and frame columns. Frame columns with weak column-strong beam conditions have a shear contribution equal to that developed when the top and bottom of the column are at flexural capacity. Where there is a strong column-weak beam condition, the column shear resistance contribution should be the shear corresponding to the development of the adjoining beam yield hinges and the column base connection capacity. In any case, the column shear contribution shall not exceed the column shear capacity.

An extreme weak story is prohibited (under §12.3.3.1) for structures more than two stories or 30 feet in height if the "weak story" has a calculated strength of less than 80 percent of the story above. A weak-story condition is absolutely prohibited in SDC E and F.

Example 9
Vertical Irregularity Type 5a   §12.3.3.1

A five-story building has a steel special moment-resisting frame (SMRF). The frame consists of W24 beams and W14 columns with the following member strength properties.

Beams at Levels 1 and 2:
  $M_{nb} = ZF_y = 250$ kip-ft
Columns on lines B and C at both levels:
  $M_{nc} = 250$ kip-ft at
  axial loading of $1.2P_D + 0.5P_L$
Column base connections at grade (based on grade-beam strength):
  $M_{nGB} = 100$ kip-ft
In addition, assume for the purposes of illustration only, that the columns have been designed such that a strong beam-weak column condition is permitted.

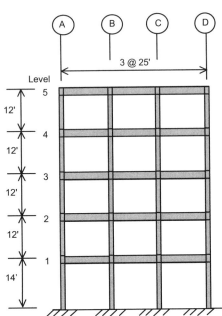

Determine if a Type 5 vertical irregularity (discontinuity in capacity-weak story) condition exists in the first story.

1.	Determine first-story strength
2.	Determine second-story strength
3.	Determine if weak-story exists at first story

Calculations and Discussion   Code Reference

A Type 5 weak-story discontinuity in capacity exists when the story strength is less than 80 percent of that of the story above (where it is less than 65 percent, an extreme weak story exists). The story strength is considered to be the total strength of all seismic-force-resisting elements that share the story shear for the direction under consideration.

To determine if a weak story exists in the first story, the sums of the column shears in the first and second stories—when the member moment capacities are developed by lateral loading—must be determined and compared.

In this example, it is assumed that the beam moments at a beam-column joint are

Example 9 ■ *Vertical Irregularity Type 5a*   *§12.3.3.1*

distributed equally to the sections of the columns directly above and below the joint. Given below are the calculations for first and second stories.

1. Determine first story strength

Columns A and D must be checked for strong column-weak beam considerations

$$2M_c = 400 > M_b = 250$$

∴ Strong column-weak beam condition exists.

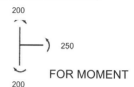

FOR MOMENT

Next, the shear in each column must be determined. Note moment capacity of beam (250/2) governs over moment capacity of column (200) to determine shear

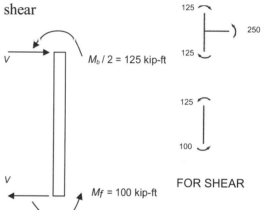

FOR SHEAR

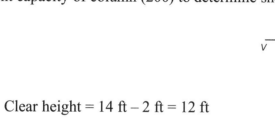

Clear height = 14 ft − 2 ft = 12 ft

$$V_A = V_D = \frac{125 + 100}{12} = 18.75 \text{ kips}$$

Checking columns B and C for strong column-weak beam considerations

$$2M_c = 400 < 2M_b = 500$$

∴ Strong beam-weak column condition exists.

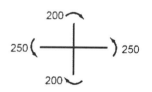

FOR MOMENT

Next, the shear in each column must be determined. Note moment capacity of column governs over moment capacity of beam to determine shear.

Clear height = 14 ft − 2 ft = 12 ft

$$V_B = V_C = \frac{200 + 100}{12} = 25.0 \text{ kips}$$

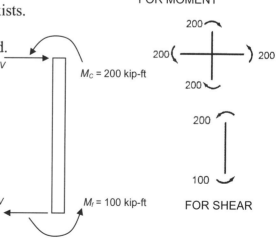

FOR SHEAR

First story strength = $V_A + V_B + V_D = 2(18.75) + 2(25.0) = 87.5$ kips

2. Determine second story strength

Columns A and D must be checked for strong column-weak beam at Level 2

$$2M_c = 400 > M_b = 250$$

∴ strong column-weak beam condition exists.

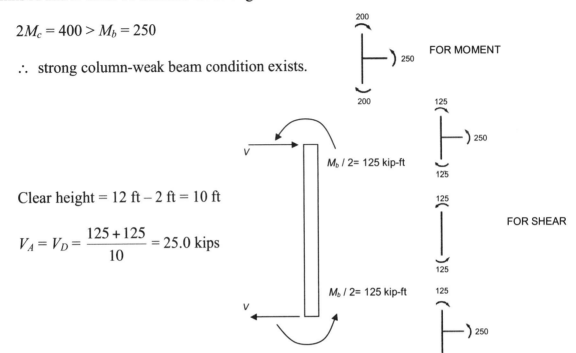

Clear height = 12 ft – 2 ft = 10 ft

$$V_A = V_D = \frac{125 + 125}{10} = 25.0 \text{ kips}$$

Checking columns B and C for strong column-weak beam considerations

$$2M_c = 400 < 2M_b = 500$$

∴ Strong beam-weak column condition exists.

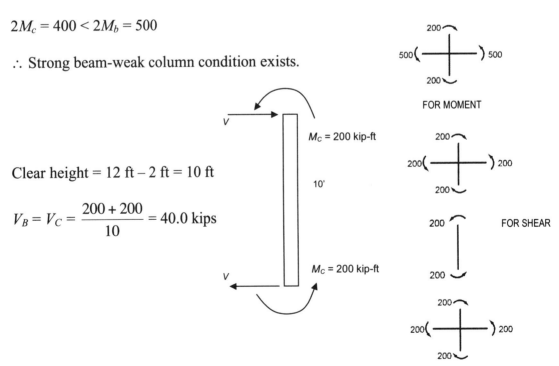

Clear height = 12 ft – 2 ft = 10 ft

$$V_B = V_C = \frac{200 + 200}{10} = 40.0 \text{ kips}$$

Second story strength
$$= V_A + V_B + V_C + V_D + 2(25.0) + 2(40.0) = 130.0 \text{ kips}$$

Example 9 ■ *Vertical Irregularity Type 5a*   *§12.3.3.1*

| **3.** | **Determine if weak story exists at first story** |

First story strength = 87.5 kips

Second story strength = 130.0 kips

$$87.5 < 0.80(130) = 104$$ (T 12.3-2, Item 5a)

∴ Weak story condition in first story exists.

Introduction to Horizontal Irregularities   §12.3.2.1

Horizontal structural irregularities are identified in Table 12.3-1. There are five types of horizontal irregularities:

1a. Torsional Irregularity — to be considered when diaphragms are not flexible as determined in §12.3.1.2

1b. Extreme Torsional Irregularity — to be considered when diaphragms are not flexible as determined in §12.3.1.2

2.   Re-entrant Corner Irregularity.

3.   Diaphragm Discontinuity Irregularity.

4.   Out-of-plane Offsets Irregularity.

5.   Nonparallel Systems – Irregularity.

These irregularities can be categorized as being either special response conditions or cases of irregular load path. Types 1a, 1b, 2, 3, and 5 are special response conditions:

Type 1a and 1b. When the ratio of maximum story drift to average story drift exceeds the given limit, there is the potential for an unbalance in the inelastic deformation demands at the two extreme sides of a story. As a consequence, the equivalent stiffness of the side having maximum deformation will be reduced, and the eccentricity between the centers of mass and rigidity will be increased along with the corresponding torsions. An amplification factor A_x is to be applied to the accidental torsion M_{ta} to represent the effects of this unbalanced stiffness, §12.8.4.1 to 12.8.4.3.

Type 2. The opening and closing deformation response or flapping action of the projecting legs of the building plan adjacent to re-entrant corners can result in concentrated forces at the corner point. Elements must be provided to transfer these forces into the diaphragms.

Type 3. Excessive openings in a diaphragm can result in a flexible diaphragm response along with force concentrations and load path deficiencies at the boundaries of the openings. Elements must be provided to transfer the forces into the diaphragm and the structural system.

Type 4. The out-of-plane offset irregularity represents the irregular load path category. In this case, shears and overturning moments must be transferred from the level above the offset to the level below the offset, and there is a horizontal offset in the load path for the shears.

Type 5. The response deformations and load patterns on a system with nonparallel lateral-force-resisting elements can have significant differences from those of a regular system. Further analysis of deformation and load behavior may be necessary.

Example 10 ▪ Horizontal Irregularity Type 1a and Type 1b   §12.3.2.1

Example 10
Horizontal Irregularity Type 1a and Type 1b   §12.3.2.1

A three-story special moment-resisting frame building has rigid floor diaphragms. Under code-prescribed seismic forces, including the effects of accidental torsion, it has the following elastic displacements δ_{xe} at Levels 1 and 2.

$$\delta_{L,2} = 1.20 \text{ in} \quad \delta_{R,2} = 1.90 \text{ in}$$

$$\delta_{L,1} = 1.00 \text{ in} \quad \delta_{R,1} = 1.20 \text{ in}$$

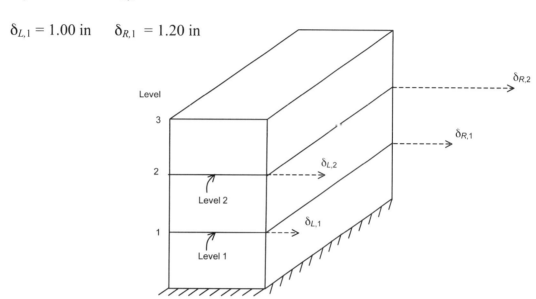

1. **Determine if a Type 1a or Type 1b torsional irregularity exists at the second story**

If it does:

2. **Compute the torsional amplification factor A_x for Level 2**

Calculations and Discussion   **Code Reference**

A Type 1a torsional irregularity is considered to exist when the maximum story drift, including accidental torsion effects, at one end of the structure transverse to an axis is more than 1.2 times the average of the story drifts of the two ends of the structure, see §12.8.6 for story drift determination

1.   Determine if a Type 1a torsional irregularity exists at the second story

Referring to the above figure showing the displacements δ_{xe} due to the
prescribed lateral forces, this irregularity check is defined in terms of story
drift $\Delta_x = (\delta_x - \delta_{x-1})$ at ends R (right) and L (left) of the structure. Torsional
irregularity exists at Level x when

T 12.3-1

$$\Delta_{max} > 1.2(\Delta_{avg}) > \frac{1.2(\Delta_{L,x} + \Delta_{R,x})}{2}$$

where

$$\Delta_{L,2} = \delta_{L,2} - \delta_{L,1}$$

$$\Delta_{R,2} = \delta_{R,2} - \delta_{R,1}$$

$$\Delta_{av} = \frac{\Delta_{L,2} + \Delta_{R,2}}{2}$$

Determining story drifts at Level 2

$$\Delta_{L,2} = 1.20 - 1.00 = 0.20 \text{ in}$$

$$\Delta_{R,2} = 1.90 - 1.20 = 0.70 \text{ in}$$

$$\Delta_{avg} = \frac{0.20 + 0.70}{2} = 0.45 \text{ in}$$

$$\Delta_{max} = 0.70 \ldots (\Delta_{R,2})$$

Checking 1.2 criteria

$$\frac{\Delta_{max}}{\Delta_{avg}} = \frac{0.7}{0.45} = 1.55 > 1.2$$

∴ Torsional irregularity exists – Type 1a.

Check for extreme torsional irregularity

$$\Delta_{max} > 1.4(\Delta_{avg})$$

$$\frac{\Delta_{max}}{\Delta_{avg}} = \frac{0.70}{0.45} = 1.55 \ldots \text{thus, extreme torsion irregularity exists – Type 1b.}$$

Example 10 ■ *Horizontal Irregularity Type 1a and Type 1b*   *§12.3.2.1*

| **2.** | **Compute amplification factor A_x for Level 2** | **§12.8.4.3** |

When torsional irregularity exists at a Level x, the accidental torsional moment M_{ta} must be increased by an amplification factor A_x. This must be done for each level, and each level may have a different A_x value. In this example, A_x is computed for Level 2.

Note that A_x is a function of the displacements as opposed to/versus the drift.

$$A_x = \left(\frac{\delta_{max}}{1.2\delta_{avg}} \right)^2 \qquad \text{(IBC Eq 16-44)}$$

$$\delta_{max} = 1.90 \text{ in} \dots (\delta_{R,2})$$

$$\delta_{avg} = \frac{\delta_{L,2} + \delta_{R,2}}{2} = \frac{1.30 + 1.90}{2} = 1.60 \text{ in}$$

$$A_2 = \left(\frac{1.90}{1.2(1.60)} \right)^2 = 0.98 < 1.0 \dots \text{Note } A_x \text{ shall not be less than 1.0}$$

$$\therefore \text{ use } A_x = 1.0.$$

Commentary

In §12.8.4.3, there is the provision that the more severe loading shall be considered. The interpretation of this for the case of the story drift and displacements to be used for the average values $\Delta\delta_{avg}$ and δ_{avg} is as follows. The most severe condition is when both $\delta_{R,X}$ and $\delta_{L,X}$ are computed for the same accidental center-of-mass displacement that causes the maximum displacement δ_{max}. For the condition shown in this example where $\delta_{R,X} = \delta_{max}$, the centers-of-mass at all levels should be displaced by the accidental eccentricity to the right side R, and both $\delta_{R,X}$ and $\delta_{L,X}$ should be evaluated for this load condition.

Table 12.3-1 triggers a number of special design requirements for torsionally irregular structures. In fact, if irregularity Type 1b (Extreme Torsional Irregularity) is present, §12.3.3.1 is triggered, which prohibits such structures for SDC E or F. It is important to recognize that torsional irregularity is defined in terms of story drift Δ_x, while the evaluation of A_x by Equation 12.8-14 is, in terms of displacements δ_{xe}. There can be instances where the story-drift values indicate torsional irregularity and where the related displacement values produce an A_x value less than 1.0. This result is not the intent of the provision, and the value of A_x used to determine accidental torsion should not be less than 1.0.

The displacement and story-drift values should be obtained by the equivalent lateral-force method with the code-prescribed lateral forces. Theoretically, if the dynamic analysis procedure were to be used, the values of Δ_{max} and Δ_{avg} would have to be found for each dynamic mode, then combined by the appropriate SRSS or CQC procedures, and then scaled to the code-prescribed base shear. However, in view of the complexity of this determination and the judgmental nature of the 1.2 factor, it is reasoned that the equivalent static force method is sufficiently accurate to detect torsional irregularity and evaluate the A_x factor.

If the dynamic analysis procedure is either elected or required, then §12.7.3 requires the use of a three-dimensional model if there are any irregularities.

For cases of large eccentricity and low torsional rigidity, the static force procedure can result in a negative displacement on one side and a positive on the other. For example, this occurs if $\delta_{L,3} = -0.40$ in. and $\delta_{R,3} = 1.80$ in. The value of δ_{avg} in Equation 12.8-14 should be calculated as the algebraic average.

$$\delta_{avg} = \frac{\delta_{L,3} + \delta_{R,3}}{2} = \frac{(-40) + 1.80}{2} = \frac{1.40}{2} = 0.70 \text{ in}$$

When dynamic analysis is used, the algebraic average value δ_{avg} should be found for each mode, and the individual modal results must be properly combined to determine the total response value for δ_{avg}.

Example 11 ▪ *Horizontal Irregularity Type 2*   *§12.3.2.1*

Example 11
Horizontal Irregularity Type 2   §12.3.2.1

The plan configuration of a ten-story special moment frame building is as shown below.

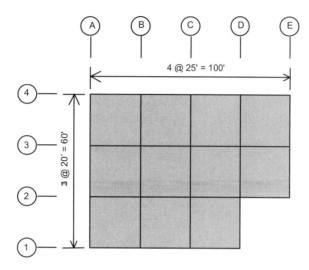

---

1.   **Determine if there is a Type 2 re-entrant corner irregularity**

Calculations and Discussion   **Code Reference**

A Type 2 re-entrant corner irregularity exists when the plan configuration of a structure and its lateral-force-resisting system contain re-entrant corners, where both projections of the structure beyond a re-entrant corner are greater than 15 percent of the plan dimension of the structure in the direction considered.

The plan configuration of this building, and its lateral-force-resisting system, has re-entrant corner dimensions as shown. For the sides on line 1, the projection beyond the re-entrant corner is

  100 ft – 75 ft = 25 ft

This is $\dfrac{25}{100}$ or 25 percent of the 100-ft plan dimension . . . More than 15 percent.

For the sides on line E, the projection is

  60 ft – 40 ft = 20 ft

This is $\frac{20}{60}$ or 33.3 percent of the 60-ft plan dimension . . . More than 15 percent.

Since both projections exceed 15 percent, there is a re-entrant corner irregularity.

∴ Re-entrant corner irregularity exists.

Commentary

Whenever the Type 2 re-entrant corner irregularity exists, see the diaphragm design requirements of §12.3.3.4 for SDC D, E, and F.

Example 12 ■ *Horizontal Irregularity Type 3*   *§12.3.2.1*

Example 12
Horizontal Irregularity Type 3   §12.3.2.1

A five-story concrete building has a bearing wall system located around the perimeter of the building. Lateral forces are resisted by the bearing walls acting as shear walls. The floor plan of the second floor of the building is shown below. The symmetrically placed open area in the diaphragm is for an atrium, and has dimensions of 40 feet by 75 feet. All diaphragms above the second floor are without significant openings.

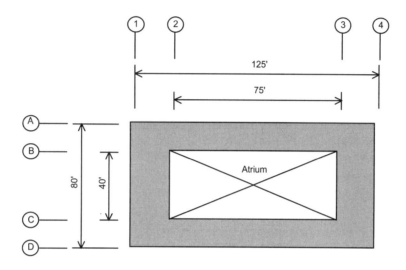

Second floor plan

1. **Determine if a Type 3 diaphragm discontinuity irregularity exists at the second floor level**

Calculations and Discussion   Code Reference

A Type 3 diaphragm discontinuity irregularity exists when diaphragms have abrupt discontinuities or variations in stiffness, including cutout or open areas comprising more than 50 percent of the gross enclosed area of the diaphragm, or changes in effective diaphragm stiffness of more than 50 percent from one story to the next.

The first check is for gross area

Gross enclosed area of the diaphragm is 80 ft × 125 ft = 10,000 sq ft

Area of opening is 40 ft × 75 ft = 3000 sq ft

50 percent of gross area = 0.5(10,000) = 5000 sq ft

3000 < 5000 sq ft

∴ No diaphragm discontinuity irregularity exists.

The second check is for stiffness.

The stiffness of the second floor diaphragm with its opening must be compared with the stiffness of the solid diaphragm at the third floor. If the change in stiffness exceeds 50 percent, a diaphragm discontinuity irregularity exists for the structure.

This comparison can be performed as follows.

Find the simple beam mid-span deflections Δ_2 and Δ_3 for the diaphragms at Levels 2 and 3, respectively, due to a common distributed load w such as 1 klf.

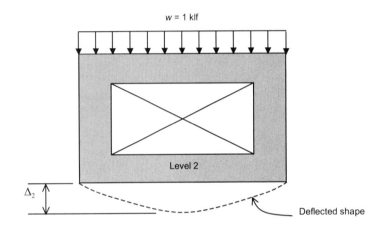

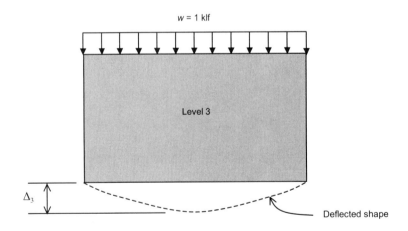

If $\Delta_2 > 1.5\Delta_3$, there is diaphragm discontinuity.

Example 13 ■ *Horizontal Irregularity Type 4*   *§12.3.2.1*

Example 13
Horizontal Irregularity Type 4   §12.3.2.1

A four-story building has a concrete shear wall lateral-force-resisting system in a building frame system configuration. The plan configuration of the shear walls is shown below.

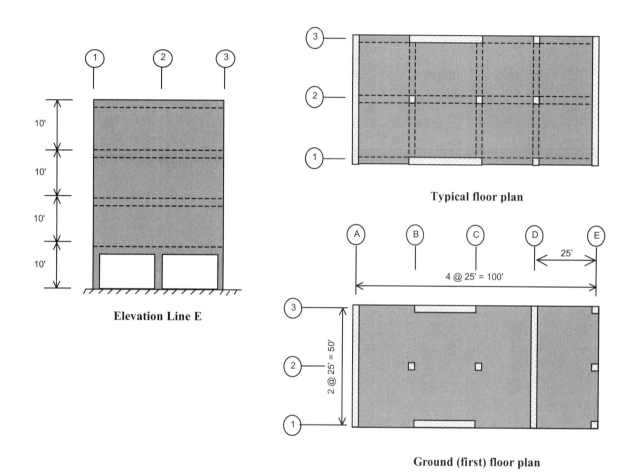

Elevation Line E

Typical floor plan

Ground (first) floor plan

---

1. Determine if there is a Type 4 out-of-plane offset irregularity between the first and second stories

---

Calculations and Discussion   **Code Reference**

An out-of-plane offset plan irregularity exists when there are discontinuities in a lateral-force path. For example: out-of-plane offsets of vertical lateral-force-resisting elements such as shear walls. The first story shear wall on line D has a 25-foot out-of-plane offset to the shear wall on line E at the second story and above. This constitutes an out-of-plane offset irregularity, and the referenced sections in Table 12.3.2.1 apply to the design.

∴ Offset irregularity exists.

Example 14
Horizontal Irregularity Type 5   §12.3.2.1

A ten-story building has the floor plan shown below at all levels. Special moment-resisting frames are located on the perimeter of the building on lines 1, 4, A, and F.

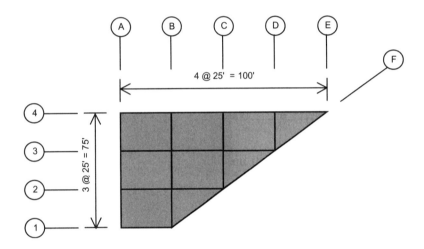

Typical floor plan

---

1. **Determine if a Type 5 nonparallel system irregularity exists**

Calculations and Discussion   **Code Reference**

A Type 5 nonparallel system irregularity is considered to exist when the vertical lateral-force-resisting elements are not parallel to or symmetric about the major orthogonal axes of the building's lateral-force-resisting system.

The vertical lateral-force-resisting frame elements located on line F are not parallel to the major orthogonal axes of the building (i.e., lines 4 and A). Therefore a nonparallel system irregularity exists, and the referenced section in Table 12.3-1 applies to the design, see §12.5.3, §12.7-3, and Table 12.6-1.

∴ A nonparallel system irregularity exists.

A 3-dimensional dynamic analysis is recommended.

Example 15 ■ *Reliability/Redundancy Coefficient ρ* §12.3.4

Example 15
Redundancy Factor ρ §12.3.4

The calculation of the redundancy factor ρ has changed considerably between earlier codes (1997 UBC; 2000 and 2003 IBC; ASCE/SEI 7-02) and the ASCE/SEI 7-05. The calculation is in some ways simpler, although it nevertheless requires some effort for conditions that do not comply with prescriptive requirements (unless the full penalty is taken, as described below).

ASCE/SEI 7-05 permits the redundancy factor to be taken as 1.0 in the following circumstances (§12.3.4.1):

1. Structures assigned to Seismic Design Category B or C. (Note that the load combinations that include the redundancy factor are not used for Seismic Design Category A.)

2. Drift calculation and P-delta effects.

3. Design of nonstructural components.

4. Design of nonbuilding structures that are not similar to buildings.

5. Design of collector elements, splices and their connections for which the load combinations with overstrength factor of §12.4.3.2 are used.

6. Design of members or connections where the load combinations with overstrength of §12.4.3.2 are required for design.

7. Diaphragm loads determined using Eq. 12.10-1 (note that this does not apply to forces transferred through a diaphragm, such as due to an out-of-plane offset in the seismic load resisting system, and the higher ρ factor may apply as otherwise required).

8. Structures with damping systems designed in accordance with 18.

Additionally, §12.3.4.2 identifies two other conditions in which ρ may be taken as 1.0. Note that the criteria for these conditions need only be met at floor levels in which more than 35-percent of the base shear is being resisted; for the top level or levels of taller structures, the conditions need not be met. The factor may be taken as 1.0 when either of the conditions listed below is met. In all other conditions, ρ is taken as 1.3. There is no longer a calculated ρ factor between the minimum and maximum values.

Condition I

12.3.4.2(a)　　Configurations in which the removal of one element (as described below in the summary of Table 12.3-3) will not result in more than a 33-percent reduction in story shear strength or in an extreme torsional irregularity (as defined in Table 12.3-1).

Summary of Table 12.3-3

Removal of one element is defined as:

1. The removal of a brace (braced frames).

2. Loss of moment resistance at the beam-to-column connections at both ends of a single beam (moment frames).

3. Removal of a shear wall or wall pier with a height-to-length ratio greater than 1.0 (shear wall systems).

4. Loss of moment resistance at the base connections of any single cantilever column (cantilever column systems).

5. For other systems, such as seismically damped structures, no prescriptive requirements are given, allowing ρ to be taken as 1.0.

Condition II

12.3.4.2(b)　　Configurations with no plan irregularities at any level and with sufficient perimeter braced frames, moment frames, or shearwalls. Sufficient perimeter bracing is defined as at least two bays of seismic force-resisting perimeter framing on each side of the structure in each orthogonal direction. For shear wall systems the number of bays is calculated as the length of shear wall divided by the story height (two times the length of shear wall divided by the story height for light-framed construction).

EXAMPLE

To illustrate the application of the method for establishing the redundancy factor, the structure shown in Figure 15.1 will be analyzed.

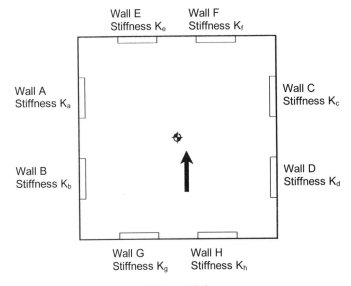

Figure 15-1

Example 15 ■ *Reliability/Redundancy Coefficient ρ* §12.3.4

Given information:

SDC D
One story, concrete shearwall building

$K_a = K_b = K_c = K_d = K_e = K_f = K_g = K_h$

All walls have the same nominal shear strength, R_n

The story height is 18 feet.

The length of each shear wall is 15 feet.

For purposes of the required strength of the walls, the redundancy factor must be determined and used in Equation 12.4-3 to determine the horizontal seismic load effect. None of the conditions listed in §12.3.4.1 apply, and thus §12.3.4.2 must be used to determine whether is 1.0 or 1.3.

Because there are two bays of shear wall on each of the perimeter lines of resistance and the building is completely regular, §12.3.4.2(b) might allow a factor of 1.0. However, the length of each shear-wall bay is less than the story height, the number of bays as defined by §12.3.4.2(b) is less than two, and thus the configuration does not automatically qualify for a redundancy factor of 1.0. The configuration will therefore be analyzed using the method outlined in §12.3.4.2(a), namely, by removing a wall and assessing the effect on story shear strength and on building torsion. In this example Wall C will be removed. Because of the symmetry of the system, the removal of one wall covers the cases of the removal of each of the other walls. In a more typical system, a separate check would need to be performed for several (or even all) of the walls.

The effect on story shear strength can be considered in at least two ways. The most conventional way to calculate the modified story shear strength is based on the modified elastic distribution of forces and the capacity of the most heavily stressed wall. Such an analysis of the structure with all four bays present shows that the seismic forces in each line of resistance (including the effects of accidental torsion) are 52.5-percent of the base shear, with each bay on each line resisting 26.25-percent; this distribution is shown in Figure 15.2(a). If the stiffness of one line of resistance is reduced by half, the design seismic forces change to 42-percent resisted on the weaker line and on the stronger line; this distribution is shown in Figure 15.2(b). Thus the increase in the force on the most heavily loaded bay is 42%/26.25% = 1.6, and the reduced force level causing yielding of that wall is 1/1.6 = 62.5%. Using this method, then, the effect on story drift is assessed to be a decrease in capacity of 100% – 62.5% = 37.5%, and thus the configuration would not qualify for a ρ factor of 1.0.

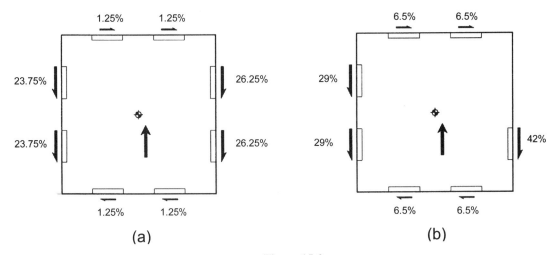

Figure 15-2

While this is an acceptable method of demonstrating compliance with the conditions justifying a factor ρ of 1.0, this method is not required. A more direct method of establishing story shear capacity is to utilize a plastic mechanism analysis. This is the method envisioned by the committee that authored the redundancy provision, and it is more consistent with the principles of seismic design (i.e., considering strength and limit states, rather than elastic design). In this method of analysis, the story shear capacity before removal of a wall is the sum of the capacities of the 4 walls resisting the seismic force in the direction under consideration (provided that the orthogonal walls have sufficient strength to resist the torsion, which in this case is only the accidental torsion). This is shown in Figure 15.3(a), where R_n denotes the capacity of the wall. If one wall is removed, the story shear capacity is the sum of the capacities of the 3 remaining walls resisting the seismic force in the direction under consideration; again, the orthogonal walls must be checked for the forces resulting from building torsion, which in this case is substantial. This is shown in Figure 15.3(b). Thus the reduction in capacity is only 25-percent. The resulting building torsional forces must be resisted by the frames in the orthogonal direction. This interpretation of the story shear capacity has been endorsed by the SEAOC Seismology Committee.

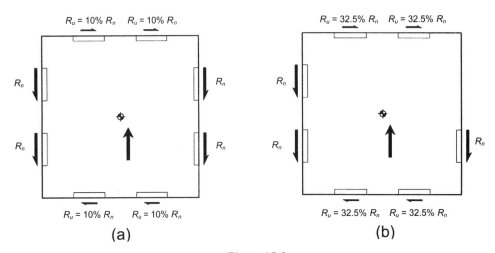

Figure 15-3

Example 15 ■ *Reliability/Redundancy Coefficient ρ*   *§12.3.4*

To qualify for a factor of 1.0, the system with one wall removed must also be checked for an extreme torsional irregularity as defined in Table 12.3-1. For the example, using the plastic mechanism analysis, the deflection in the direction of loading is R_n/K_a. The additional deflection at each perimeter line due to rotation is $0.325R_n/K_a$. This is less than the 40-percent maximum that is allowed by Table 12.3-1 before an extreme torsional irregularity is deemed to exist. Thus, the configuration qualifies for a ρ factor of 1.0.

Example 16
P-delta Effects   §12.8.7

In high-rise building design, important secondary moments and additional story drifts can be developed in the lateral-force-resisting system by *P*-delta effects. *P*-delta effects are the result of the axial load *P* in a column being moved laterally by horizontal displacements, thereby causing additional secondary column and girder moments. The purpose of this example is to illustrate the procedure that must be used to check the overall stability of the frame system for such effects.

A 15-story building has a steel special moment frame (SMF).

The following information is given.

Seismic Use Group I

Seismic Design Category D

$R \; = 8$

$C_d \; = 5.5$

$I \; = 1.0$

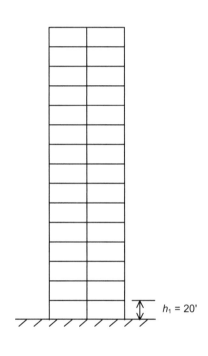

At the first story,
$\quad \Sigma D \; = W = 8643$ kips
$\quad \Sigma L \; = 3850$ kips
$\quad V_1 \; = V = 0.042W = 363.0$ kips, $\beta = 0.80$
$\quad h_1 \; = 20$ ft
$\quad$ Deflection at level $x = 1$ due to seismic base shear V (without *P*-delta effects)
$\quad \delta_{1e} \; = 0.003h_1 = 0.72$ in

Determine the following.

1.	**Initial design story drift** Δ **in first story**
2.	***P*-delta criteria for the building**
3.	**Check the first story for *P*-delta requirements**
4.	**Final design story drift and story shear in first story**
5.	**Check for story drift compliance in first story**

Example 16 ▪ *P-delta Effects*   *§12.8.7*

Calculations and Discussion	Code Reference

1. **Initial design story drift Δ in first story**   §12.8.6

At story $x = 1$, the preliminary design story drift is

$$\Delta = (\delta_x - \delta_{x-1}) = (\delta_1 - \delta_0) = (\delta_1 - 0) = \delta_1$$

where

$$\delta_1 = \frac{C_d \delta_{1e}}{I} = \frac{5.5(0.72)}{1.0} = 3.96 \text{ in} \qquad \text{(Eq 12.8-15)}$$

Now : $\Delta = 3.96$

This value is termed initial because it may need to be increased by the incremental factor $a_d = 1.0/(1-\theta)$ as determined in Part $\boxed{3}$ of this example.

2. **P-delta criteria for the building**   §12.8.7

P-delta effects must be considered whenever the ratio of secondary moments to primary moments exceeds 10 percent. This ratio is defined as stability coefficient θ

$$\theta = \frac{P_x \Delta}{V_x h_{sx} C_d} \qquad \text{(Eq 12.8-16)}$$

where

   θ   = stability coefficient for story x

   P_x = total design vertical load on all columns in story x
        (Note: no factor above 1.0 is required)

   Δ   = initial design story drift in story x occurring simultaneously with $C_d V_x$

   V_x = seismic shear force in story x

   h_{sx} = height of story x

   C_d = deflection amplification factor in Table 12.2-1 (given = 5.5)

   P-delta effects must be considered when $\theta > 0.10$

3.   **Check *P*-delta requirements for the first story**   **§12.8.7**

Section 12.8.7 requires that the total vertical load P_1 at the first story be considered the total dead ΣD plus floor live ΣL and snow load S above the first story. These loads are unfactored for determination of *P*-delta effects.

$$P_1 = \Sigma D + \Sigma L + S$$

Using $S = 0$ for the building site,

$$P_1 = 8643 + 3850 = 12{,}493 \text{ kips}$$

For story $x = 1$,

$$\theta_1 = \frac{P_1 \Delta}{V_1 h_{s1} C_d} = \frac{(8643 + 3850)(3.96)}{(363.0)(20 \text{ ft})(12)(5.5)} = 0.103 > 0.100$$

$\therefore$ *P*-delta effects must be considered.

Check for $\theta \leq \theta_{max}$ using the given $\beta = 0.80$

$$\theta_{max} = \frac{0.5}{\beta C_d} = \frac{0.5}{(0.80)(5.5)} = 0.1136 \qquad \text{(Eq 12.8-17)}$$

$$0.103 < 0.1136 \ldots o.k.$$

4.   **Final design story drift and story shear in first story**   **§12.8.7**

When $\theta > 0.10$, the initial design story drift and design story shear must be augmented by the incremental factor a_d related to *P*-delta effects

$$a_d = \frac{1.0}{1 - \theta} = \frac{1.0}{1 - 0.103} = 1.115$$

The final design story drift in the first story is

$$\Delta_1 = a_d \Delta = (1.115)(3.96) = 4.415 \text{ in}$$

The final design story shear is

$$V_1 = a_d V_1 = (1.115)(363.0) = 404.7 \text{ kips}$$

Example 16 ■ *P-delta Effects* §12.8.7

| **5.** | **Check for story-drift compliance in the first story** | **§12.8.7** |

Allowable story drift $\Delta_{allow} = 0.020\ h_1$ T 12.12-1

$$\Delta_{allow} = 0.020(20\text{ ft})(12) = 4.80\text{ in}$$

$$\Delta_1' = 4.415 < 4.80\text{ in} \dots o.k.$$

Commentary

In §12.8.7 the *P*-delta effects on the design story drift and the design story shear are evaluated by the following procedure:

1. Given the initial design story drift $\Delta_x = \delta_x - \delta_{x-1}$ at story x: compute for each story x the stability coefficient θ_x given by Equation 12.8-16. For each story where θ_x is equal to, or greater than 0.10, compute the corresponding incremental factor relating to *P*-delta effects $a_d = 1/(1 - \theta_x)$. This factor accounts for the multiplier effect due to the initial story drift Δ_x leading to another increment of story drift, leading to another story drift, which would lead to yet another increment, etc. Thus both the drift and the shear in the story would be increased by a factor equal to the series of $1 + \theta + \theta^2 + \theta^3 + \text{---}$, which converges to $1(1 - \theta) = a_d$. As a result the initial story drift Δ_x and story shear V_x need to be multiplied by the factor a_d to represent the total final *P*-delta effect.

2. The final resulting story drift $\Delta_x' = a_d\,\Delta_x$ needs to comply with the drift limitations of §12.12.

3. In each story requiring consideration of *P*-delta effects the initial story shears are increased to $V_x' = a_d V_x$. The structural elements must be designed to resist the resulting final story shears, overturning moments and element actions.

Some computer programs for frame analysis state that *P*-delta effects are included directly in the analysis. The engineer should verify that the total gravity load employed and the method used in these programs will provide results that are essentially equivalent to the augmented story shear method described above.

The provisions in §§12.8.6 and 12.8.7 for the evaluation of the final story drifts state that the final story drift shall be a_d times the initial drift Δ.
However, in a multi-story building having $\theta > 0.1$ in more than one story, the initial story shears in these stories are increased by the a_d factor. This is equivalent to an added lateral load equal to $(a_d - 1)\,V_x$ applied to each story level having $\theta > 0.1$. Therefore the new story drifts in the stories below would be increased not only by their own a_d but by the added lateral load effect from the stories above; thus, the final drifts should be found by a new analysis with the added lateral loads equal to $(a_d - 1)\,V_x$ along with the initial lateral loads on the frame.

Example 17
Seismic Base Shear   §12.8.1

Find the design base shear for a 5-story steel special moment-resisting frame building shown below.

The following information is given.

Seismic Design Category D

$$S_{DS} = 0.45g$$
$$S_{D1} = 0.28g$$
$$I = 1.0$$
$$R = 8$$
$$W = 1626 \text{ kips}$$
$$h_n = 60 \text{ feet}$$

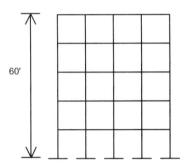

60'

To solve this example, follow these steps.

1. **Determine the structure period**

2. **Determine the seismic response coefficient C_s**

3. **Determine seismic base shear**

Calculations and Discussion   **Code Reference**

1. **Determine the structure period**   **§12.8.2.1**

The appropriate fundamental period T_a is to be used. C_T for steel moment-resisting frames is 0.028.

$$T_a = C_T(h_n)^{3/4} = 0.028(60)^{0.8} = 0.74 \text{ sec} \qquad \text{(Eq 12.8-7)}$$

2. **Determine the seismic response coefficient C_s**   **§12.8.1**

The design value of C_s is the smaller value of

$$C_s = \frac{S_{DS}}{\left(\dfrac{R}{I}\right)} = \frac{(0.45)}{\left(\dfrac{8}{1.0}\right)} = 0.0562 \qquad \text{(Eq 12.8-2)}$$

Example 17 ■ *Seismic Base Shear* **§12.8.1**

and

$$C_s = \frac{S_{D1}}{T\left(\dfrac{R}{I}\right)} = \frac{(0.28)}{\left(\dfrac{8}{1.0}\right)(0.74)} = 0.047 \text{ for } T \le T_L \qquad \text{(Eq 12.8-3)}$$

$$C_s = \frac{S_{D1}T_L}{T^2\left(\dfrac{R}{I}\right)} = \text{ for } T > T_L \qquad \text{(Eq 12.8-4)}$$

but shall not be less than

$$C_s = 0.01 \qquad \text{(Eq 12.8-5)}$$

In addition, for structures located where S_1 is equal to or greater than $0.6g$, C_s shall not be less than

$$C_s = \frac{0.5S_1}{\left(\dfrac{R}{I}\right)} \qquad \text{(Eq 12.8-6)}$$

∴ Design value of $C_s = 0.0467$

3. Determine seismic base shear §12.8.1

The seismic base shear is given by

$$V = C_sW \qquad \text{(Eq 12.8-1)}$$

$$= 0.047(1626 \text{ kips})$$

$$= 76.4 \text{ kips}$$

Commentary

The S_{D1} value of $0.28g$ given in this example is based on an S_1 value of $0.21g$. If the S_1 value were to have been equal or greater than $0.6g$, then the lower bound on C_s is

$$C_s \ge \frac{0.5IS_1}{R} \qquad \text{(Eq 12.8-6)}$$

Example 18
Approximate Fundamental Period   §12.8.2.1

Determine the period for each of the structures shown below using the appropriate fundamental period formula

$$T_a = C_t(h_n)^x$$ (Eq 12.8-7)

The coefficient C_T and the exponent x are dependent on the type of structural system used.

1.	Steel special moment frame (SMF) structure
2.	Concrete special moment frame (SMF) structure
3.	Steel eccentric braced frame (EBF)
4.	Masonry shear wall building
5.	Tilt-up building

Calculations and Discussion   Code Reference

1. Steel special moment frame (SMF) structure   §12.8.2.1

Height of the structure above its base is 96 feet. The additional 22-foot depth of the basement is not considered in determining h_n for period calculation.

$$C_T = 0.028; \ x = 0.8$$

$$T_a = C_T(h_n)^x = 0.028(96)^{0.8} = 1.08 \text{ sec}$$

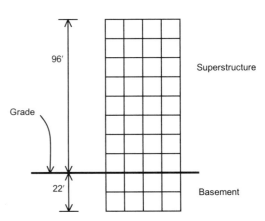

Note: In the SEAOC Blue Book, base is defined as the level at which earthquake motions are considered to be imparted, or the level at which the structure, as a dynamic vibrator, is supported. For this structure the solution is the same.

Example 18 ■ *Approximate Fundamental Period* §12.8.2.1

2. Concrete special moment frame (SMF) structure

Height of the tallest part of the building is 33 feet, and this is used to determine period. Roof penthouses are generally not considered in determining h_n, but heights of setbacks are included. However, if the setback represents more than a 130-percent change in the lateral force system dimension, there is a vertical geometric irregularity (Table 12.3-2) and dynamic analysis is required for this type of irregularity per Table 12.6-1.

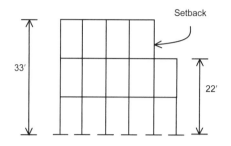

$h_n = 33$ feet

$C_T = 0.016; x = 0.9$

$T_a = C_T(h_n)^x = 0.016(33)^{0.9} = 0.37$ sec

3. Steel eccentric braced frame (EBF)

EBF structures use the $C_t = 0.03$ and $x = 0.75$.

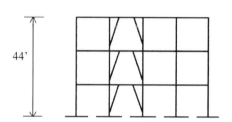

$C_T = 0.030; x = 0.75$

$T = C_T(h_n)^x = 0.030(44)^{0.75} = 0.51$ sec

4. Masonry shear wall building

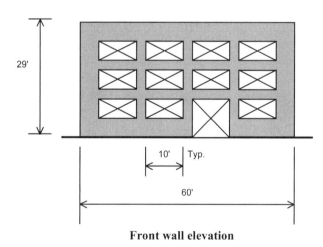

Front wall elevation

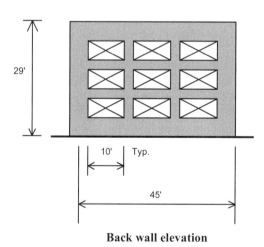

Back wall elevation

For this structure, C_T may be taken as 0.020 and x may be taken as 0.75, the values for "all other buildings"

$$T_a = C_T(h_n)^x = 0.020(29)^{0.75} = 0.25 \text{ sec}$$

5.   Tilt-up building

Consider a tilt-up building 150 feet by 200 feet in plan that has a panelized wood roof and the typical wall elevation shown below.

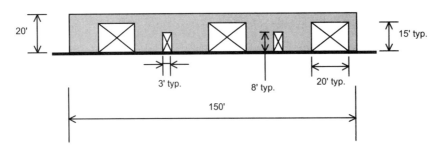

Typical wall elevation

$$C_T = 0.020; \; x = 0.75$$

$$T = C_T(h_n)^x = 0.020(20)^{0.75} = 0.19 \text{ sec}$$

This type of structural system has relatively rigid walls and a flexible roof diaphragm. The code formula for period does not take into consideration the fact that the real period of the building is highly dependent on the roof diaphragm construction. Thus, the period computed above is not a good estimate of the real fundamental period of this type of building. It is acceptable, however, for use in determining design base shear.

Commentary

The fundamental period T of the building may also be established by analytical procedures with the limitation given in §12.8.2.

Example 19 ■ *Simplified Alternative Structural Design Procedure* §*12.14*

Example 19
Simplified Alternative Structural Design Procedure §12.14

Determine the seismic base shear and the seismic lateral forces for a three-story wood structural panel wall building using the simplified alternative structural design procedure.

The following information is given.

Occupancy Category I

$$S_{DS} = 1.0$$

$$R = 6\frac{1}{2}$$

$$W = 750 \text{ kips}$$

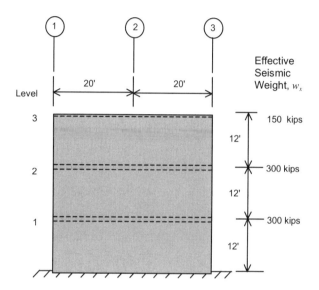

To solve this example, follow these steps.

1. Check applicability of simplified alternative method

2. Determine seismic base shear

3. Determine seismic lateral forces at each level

Calculations and Discussion   Code Reference

1. Check applicability of simplified alternative method   §12.14.1.1

Light-framed construction not more than three stories, or other buildings with bearing walls or building frame systems not more than three stories, can use the simplified alternative method when general conditions are satisfied.

| 2. | **Determine seismic base shear** | **§12.14.7.1** |

$$V = \frac{FS_{DS}W}{R} \qquad \text{(Eq 12.14-9)}$$

$$= \frac{1.2(1.0)(750 \text{ kips})}{6\frac{1}{2}}$$

$$= 138.5 \text{ kips}$$

| 3. | **Determine seismic lateral forces at each level** | **§12.14.7.2** |

$$F_x = \frac{W_x}{W}\,V \qquad \text{(Eq 12.14-10)}$$

$$F_1 = \frac{300}{750}\,(138.5) = 55.4 \text{ kips}$$

$$F_2 = \frac{300}{750}\,(138.5) = 55.4 \text{ kips}$$

$$F_3 = \frac{150}{750}\,(138.5) = 27.7 \text{ kips}$$

Commentary   **§12.8**

The following is a comparison of simplified base shear with standard design base shear. The standard method of determining the seismic base shear is

$$V = C_S W \qquad \text{(Eq 12.8-1)}$$

where

$$C_S = \frac{S_{DS}}{\left(\dfrac{R}{I}\right)} \qquad \text{(Eq 12.8-2)}$$

$$V = \frac{(1.0)(750)}{\left(\dfrac{6\frac{1}{2}}{1.0}\right)} = 115.4 \text{ kips}$$

Example 19 ■ *Simplified Alternative Structural Design Procedure*   *§12.14*

The distribution of seismic forces over the height of the structure is

$$F_x = C_{vx}V$$ (Eq 12.8-11)

where

$$\frac{w_x h_x^k}{\sum\limits_{i=1}^{n} w_i h_i^k}$$ (IBC Eq 16-42)

Note: distribution exponent $k = 1.0$ for structures having a period of 0.5 second or less.

Level x	h_x	w_x	$w_x h_x$	$\dfrac{w_x h_x}{\sum w_i h_i}$	F_x	F_x/w_x
3	36 ft	150 kips	5,400 kip-ft	0.333	38.4 kips	0.278
2	24	300	7,200	0.444	51.2	0.185
1	12	300	3,600	0.222	25.6	0.093
			$\sum w_i h_i$ 16,200		$\Sigma = 115.4$	

The seismic base shear V and lateral forces F_x at each level except the roof are all less than the simplified method, see table below. The principal advantage of the simplified method is that period T need not be calculated and design story drift Δ may be taken as 1 percent of the story height, §12.14.7.5.

Comparison of Simplified VS Standard

Level x	Lateral Force F_x		% Difference
	Simplified	Standard	
3	27.7 kips	38.4 kips	72
2	55.4	51.2	108
1	55.4	25.6	216
Total	138.5	115.2	120

Example 20
Combination of Structural Systems: Vertical   §12.2.3.1

In structural engineering practice, it is sometimes necessary to design buildings that have a vertical combination of different lateral-force-resisting systems. For example, the bottom part of the structure may be a rigid frame and the top part may be a braced frame or shear wall. This example illustrates use of the requirements of §12.2.3.1 to determine the applicable response modification coefficient R system overstrength factor Ω_o and deflection amplification factor C_d values for combined vertical systems.

For the three systems shown below, determine the required R coefficient, Ω_o factor, C_d factor, and related design base shear requirements.

Calculations and Discussion   Code Reference

1. **Steel Special concentrically braced frame (SCBF) over steel special moment frame (SMF)**

Seismic Design Category C

Ordinary steel concentrically braced frame

$R = 6.0$
$\Omega_o = 2.0$
$C_d = 5.0$
$\rho = 1.0$

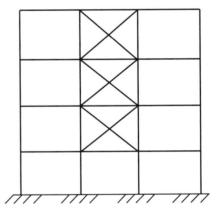

Special steel moment frame

$R = 8.0$
$\Omega_o = 3.0$
$C_d = 5.5$
$\rho = 1.0$

This combined system falls under vertical combinations of §12.2.3.1. Because the rigid framing system is above the flexible framing system, the exception for a two-stage analysis in §12.2.3.1 cannot be used. Therefore, the structure in this direction must use the lowest $R = 6.0$ and the largest $\Omega_o = 3.0$. Recall that <u>if</u> the floor and roof diaphragms could be considered to be flexible, Ω_o would be 2.5, per footnote g, Table 12.2-1.

Example 20 ▪ *Combination of Structural Systems: Vertical* §12.2.3.1

2. Ordinary reinforced concrete shear wall (ORCSW) over special reinforced concrete moment frame (SRCMF)

Seismic Design Category C

Ordinary reinforced concrete
shear wall (non-bearing)

R = 5
Ω_o = 2.5
C_d = 4.5
ρ = 1.0

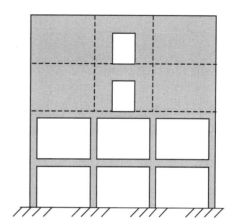

Special reinforced concrete
moment frame

R = 8.0
Ω_o = 3.0
C_d = 5.5
ρ = 1.0

This combined system falls under vertical combinations of §12.2.3.1. Because the rigid portion is above the flexible portion, a two-stage analysis cannot be used. Therefore, the structure in this direction must use the lowest, R = 5.0, and the largest, Ω_o = 3.0. Also note that ordinary reinforced concrete shear wall systems are not permitted above 35 feet in SDC D, E, or F (Table 12.2-1).

3. Concrete SRCMF over a concrete building frame system

a. Applicable criteria.

This is a vertical combination of a flexible system over a more rigid system. Under §12.2.3.1, a two-stage static analysis may be used, provided the structures conform to the following four requirements.

Seismic Design Category B

Concrete special reinforced concrete
moment-frame

R = 8.0
Ω_o = 3.0
C_d = 5.5
ρ = 1.3
Stiffness upper portion = 175 kip-in
T_{upper} = 0.55 sec
$T_{combined}$ = 0.56 sec

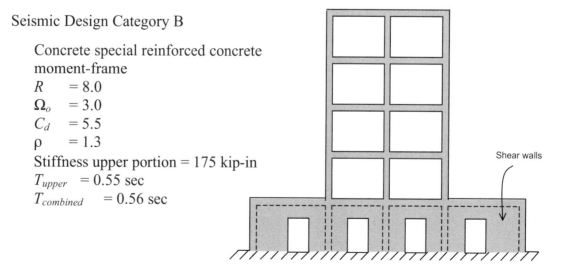

Shear walls

Special reinforced concrete shear wall

$R = 6.0$
$\Omega_o = 2.5$
$C_d = 5$
$\rho = 1.0$
Stiffness = 10,000 kip-in
$T_{lower} = 0.03$ sec

Also note R is different for <u>bearing wall systems</u> versus <u>building frame systems</u> for special reinforced concrete shear walls, see Table 12.2-1.

Check requirements of §12.2.3.1 for a two-stage analysis

a. The stiffness of the lower portion is at least 10 times the stiffness of the upper portion. For multi-story upper or lower portions, the stiffness should be the stiffness of the first mode.

    10,000 kip-in > 10(175) = 1750 kip-in . . . *o.k.*

b. Period of entire structure is not greater than 1.1 times the period of upper structure considered a separate structure fixed at the base.

    0.56 sec < 1.1 (.55) = 0.61 sec . . . *o.k.*

c. Flexible upper portion supported on the rigid lower portion shall be designed as a separate structure using appropriate values of R and ρ.

d. Rigid lower portion shall be designed as a separate structure using appropriate values of R and ρ. Reactions from the upper structure shall be determined from analysis of the upper structure amplified by the ratio of R/ρ of the upper structure over R/ρ of the lower structure. This ratio shall not be less than 1.0.

Example 20 ▪ *Combination of Structural Systems: Vertical*   *§12.2.3.1*

b. Design procedures for upper and lower structures

Design upper SRCMF using

$$R = 8.0$$
$$\Omega = 3.0$$
$$\rho = 1.3$$

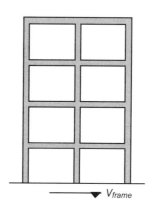

Design lower portion of the building frame system for the combined effects of amplified reactions from the upper portion and lateral forces due to the base shear for the lower portion of the structure (using $R = 6.0$, $\Omega = 2.5$, and $\rho = 1.0$ for the lower portion).

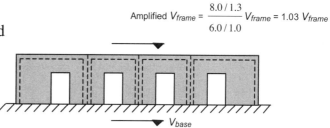

Amplified $V_{frame} = \dfrac{8.0/1.3}{6.0/1.0} V_{frame} = 1.03 \, V_{frame}$

$$\therefore V_{base} = (\text{amplified } V_{frame}) + (V_{lower}).$$

The reactions from the upper portion shall be determined from the analysis of the upper portion amplified by the ratio of (R/ρ) for the upper portion over (R/ρ) of the lower portion.

Note that for the basic seismic load combinations the factor ρ must still be applied to forces corresponding to V_{lower}.

Example 21
Combination of Framing Systems in Different Directions   §12.2.2

This example illustrates the determination of response modification coefficient R, system over strength factor Ω_o , and deflection amplification factor C_d values for a building that has different seismic framing systems along different axes (i.e., directions) of the building.

In this example, a three-story building has concrete shear walls in one direction and concrete moment frames in the other. Floors are concrete slab, and the building is SDC D and Occupancy Category II.

Determine the R, C_d, and Ω_o values for each direction.

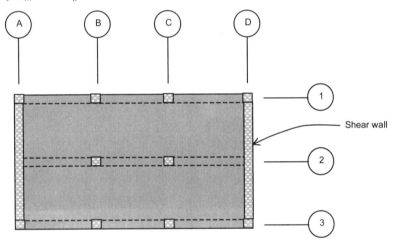

Typical Floor Plan

Lines A and D are special reinforced concrete shear walls (bearing wall system)
$R = 5$, $\Omega_o = 2.5$, $C_d = 5$, Table 12.2-1 (A1)

Lines 1, 2, and 3 are special reinforced concrete moment frames
$R = 8.0$, $\Omega_o = 3.0$, $C_d = 5.5$ Table 12.2-1 (C5)

| 1. | **Determine the *R* value for each direction** |

Calculations and Discussion

Code Reference

The provisions of §12.2.2 require that where different seismic-force-resisting systems are used along the two orthogonal axes of the structure, the appropriate response modification coefficient R, system overstrength factor Ω_o, and deflection amplification factor C_d for each system shall be used.

Use $R = 5.0$, $\Omega_o = 2.5$, and $C_d = 5$ for the north-south direction.

and $R = 8.0$, $\Omega_o = 3.0$, and $C_d = 5.5$ for the east-west direction.

Commentary

Note that since this is SDC D, ordinary reinforced concrete shear walls are not permitted.

Example 22
Combination of Structural Systems:
Along the Same Axis   §12.2.3.2

Occasionally, it is necessary or convenient to have different structural systems in the same direction. This example shows how the response modification coefficient R value is determined in such a situation.

A one-story steel frame structure has the roof plan shown below. The structure is assigned to Occupancy Category II.

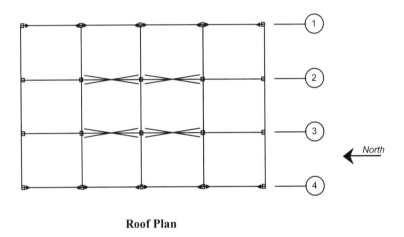

Roof Plan

Lines 1 and 4 are ordinary steel moment frames: $R = 3.5$

Lines 2 and 3 are special steel concentrically braced frames: $R = 6.0$

| 1. | **Determine the R value for the N/S direction** |

Calculations and Discussion   **Code Reference**

When a combination of structural systems is used in the same direction, §12.2.3.2 requires that (except for dual systems and shear wall-frame interactive systems) the value of R used shall not be greater than the least value of any system utilized in that direction.

∴ Use $R = 3.5$ for entire structure.

Commentary

An exception is given for light frame, <u>flexible</u> diaphragm buildings of Occupancy Category I or II two stories or less in height. However, to qualify as a flexible diaphragm, the lateral deformation of the diaphragm must be more than two times the average story drift of the associated story; see definition in §12.3.1.3.

Example 23 ■ *Vertical Distribution of Seismic Force* §*12.8.3*

Example 23
Vertical Distribution of Seismic Force §12.8.3

A nine-story building has a moment-resisting steel frame for a lateral-force-resisting system. Find the vertical distribution of lateral forces F_x.

The following information is given.

$W = 3762$ kips
$C_s = 0.062$
$R = 8.0$
$\Omega_o = 3.0$
$I = 1.0$
$T = 1.06$ sec

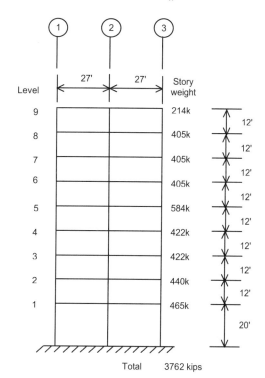

To solve this example, follow these steps.

1.	Determine V
2.	Find F_x at each level
3.	Find the distribution exponent k
4.	Determine vertical force distribution

Calculations and Discussion

Code Reference

1. Determine V §12.8-1

This is the total design lateral force or shear at the base of the structure. It is determined as follows

$$V = C_s W = 0.062\,(3762^k) = 233.8 \text{ kips} \qquad \text{(Eq 12.8-1)}$$

2. **Find F_x at each level**

The vertical distribution of seismic forces is determined as

$$F_x = C_{vx} \, V \qquad \text{(Eq 12.8-11)}$$

where

$$C_{vx} = \frac{W_x h_x^k}{\displaystyle\sum_{i=1}^{n} W_i \, h_i^k} \qquad \text{(Eq 12.8-12)}$$

Since there are nine levels above the ground, $n = 9$

Thus:

$$F_x = \frac{233.2 w_x h_x^k}{\displaystyle\sum_{i=1}^{9} w_i h_i^k}$$

3. **Find the distribution exponent k**   §12.8.3

The distribution exponent k is equal to 1.0 for buildings having a period of $T \le 0.5$ seconds, and is equal to 2.0 for buildings having a period of $T \ge 2.5$. For intermediate value of the building period, k is determined by linear interpolation.

Thus:

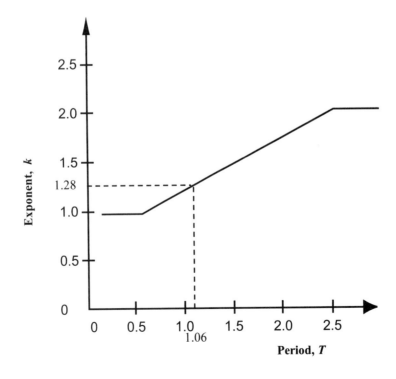

Example 23 ▪ Vertical Distribution of Seismic Force §12.8.3

Now:

for $T = 1.06$ sec

$$k = 1.0 + (1.06 - 0.5)\left(\frac{1}{2.5 - 0.5}\right)$$

$$= 1.28$$

Use: $k = 1.28$

---

4. **Equation 12.8-12 is solved in the table below given V = 233.8 kips and k = 1.28**

Level x	h_x	h_x^k (ft)	w_x (kips)	$w_x h_x^k$ kip-ft	$C_{vx} = \dfrac{w_x h_x^k}{\Sigma w_i h_i^k}$	$F_x = C_{vx} V$ (kips)	$F_x/w_x = \overline{S_a}$
9	116 ft	439	214	93,946	0.116	27.3	0.127
8	104	382	405	154,710	0.192	44.8	0.111
7	92	326	405	132,030	0.169	38.3	0.094
6	80	273	405	110,565	0.137	32.1	0.079
5	68	222	584	129,648	0.161	37.6	0.064
4	56	173	422	73,006	0.091	21.2	0.050
3	44	127	422	53,594	0.067	15.5	0.037
2	32	84	440	36,960	0.046	10.8	0.024
1	20	46	465	21,390	0.027	6.2	0.013
			$\Sigma = 3762$	$\Sigma = 806,289$	1.004	233.2	

Commentary

Note that certain types of vertical irregularity can result in a dynamic response having a load distribution significantly different from that given in this section. Table 12.6-1 lists the minimum allowable analysis procedures for seismic design. Redundancy requirements must also be evaluated once the type of lateral-force-resisting system to be used is specified, because this may require modification of the building framing system and vertical distribution of horizontal forces as a result of changes in building period T.

Often, the horizontal forces at each floor level are increased when ρ is greater than 1.0. This is done to simplify the analysis of the framing members. The horizontal forces need not be increased at each floor level when ρ is greater than 1.0, provided that, when designing the individual members of the lateral-force-resisting system, the seismic forces are factored by ρ. When checking building drift, ρ = 1.0 (§12.3.4.1) shall be used.

Structures that have a vertical irregularity of Type 1a, 1b, 2, or 3 in Table 12.3-2, or horizontal irregularities of Type 1a or 1b in Table 12.3-1, may have significantly different force distributions. Structures with long periods, e.g., $T > 3.5T_s$, require a dynamic analysis per Table 12.6-1 in Seismic Design Categories D, E, or F. In addition, some Irregular Structures require a dynamic analysis per Table 12.6-1. The configuration and final design of this structure must be checked for irregularities. Most structural analysis programs used today perform this calculation, and it is rarely necessary to manually perform the calculations shown above. However, it is recommended that these calculations be performed to confirm the computer analysis and to gain insight to structural behavior. Note that $(\overline{S_a})_{max}$ is approximately twice C_s, and $\overline{S_a} = \Gamma \phi S_a$ from a modal analysis.

Example 24 ■ *Horizontal Distribution of Shear* §12.8.4

Example 24
Horizontal Distribution of Shear §12.8.4

A single-story building has a rigid roof diaphragm. See appendix to this example for a procedure for the distribution of lateral forces in structures with rigid diaphragms and cross walls and/or frames of any orientation. Lateral forces in both directions are resisted by shear walls. The mass of the roof can be considered to be uniformly distributed, and in this example, the weight of the walls is neglected. In actual practice, particularly with concrete shear walls, the weight of the walls should be included in the determination of the center-of-mass (CM).

The following information is given.

Design base shear: $V = 100$ kips in north-south direction

Wall rigidities: $R_A = 300$ kip/in
$R_B = 100$ kip/in
$R_C = R_D = 200$ kip/in
Center-of-mass: $x_m = 40$ ft, $y_m = 20$ ft

Analyze for seismic forces in north-south direction.

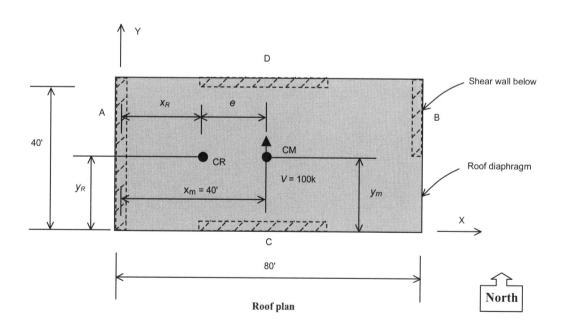

Roof plan

Determine the following.

1. Eccentricity and rigidity properties

2. Direct shear in walls A and B

3. Plan irregularity requirements

4. Torsional shear in walls A and B

5. Total shear in walls A and B

Calculations and Discussion   *Code Reference*

1. **Eccentricity and rigidity properties**   *§12.8.4.1*

The rigidity of the structure in the direction of applied force is the sum of the rigidities of walls parallel to this force.

$$R = R_A + R_B = 300 + 100 = 400 \text{ kip/in}$$

The centers of rigidity (CR) along the x and y axes are

$$x_R = \frac{R_B(80 \text{ ft})}{R_A + R_B} = 20 \text{ ft}$$

$$y_R = \frac{R_C(40 \text{ ft})}{R_C + R_D} = 20 \text{ ft}$$

eccentricity $e = x_m - x_R = 40 - 20 = 20 \text{ ft}$

Torsional rigidity about the center of rigidity is determined as

$$J = R_A (20)^2 + R_B (60)^2 + R_C (20)^2 + R_D (20)^2$$

$$= 300 (20)^2 + 100 (60)^2 + 200 (20)^2 + 200 (20)^2 = 64 \times 10^4 \text{ (kip/in) ft}^2$$

The seismic force V applied at the CM is equivalent to having V applied at the CR together with a counter-clockwise torsion T. With the requirements for accidental eccentricity e_{acc}, the total shear on walls A and B can be found by the addition of the direct and torsional load cases.

Example 24 ■ *Horizontal Distribution of Shear*   §*12.8.4*

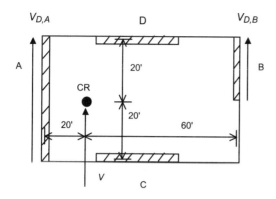

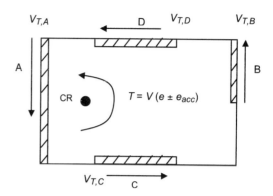

Direct shear contribution   Torsional shear contribution

2. Direct shear in walls A and B

$$V_{D,A} = \frac{R_A}{R_A + R_B} \times (V) = \frac{300}{300 + 100} \times 100 = 75.0 \text{ kips}$$

$$V_{D,B} = \frac{R_B}{R_A + R_B} \times (V) = \frac{100}{300 + 100} \times 100 = 25.0 \text{ kips}$$

3. Plan irregularity requirements

The determination of torsional irregularity, Items 1a and 1b in Table 12.3-1, requires the evaluation of the story drifts in walls A and B. This evaluation must include accidental torsion caused by an eccentricity of 5 percent of the building dimension.

$$e_{acc} + 0.05 \, (80 \text{ ft}) = 4.0 \text{ ft}$$

For the determination of torsional irregularity, the initial most severe torsional shears, V' and corresponding story drifts (so as to produce the lowest value of the average story drift) will result from the largest eccentricity $e + e_{acc}$. These are

$$V'_{T,A} = \frac{V(e + e_{acc})(x_R)(R_A)}{J} = \frac{100(20 + 4)(20)(300)}{64 \times 10^4} = 22.5 \text{ kips}$$

$$V'_{T,B} = \frac{V(e + e_{acc})(80 - x_R)(R_B)}{J} = \frac{100(20 + 4)(60)(100)}{64 \times 10^4} = 22.5 \text{ kips}$$

The initial total shears are

$$V'_A = V_{D,A} - V'_{T,A} = 75.0 - 22.5 = 52.5 \text{ kips}$$

$$V'_B = V_{D,B} + V'_{T,B} = 25.0 + 22.5 = 47.5 \text{ kips}$$

(NOTE: This is not the design force for Wall A, as accidental eccentricity here is used to reduce the force).

The resulting displacements δ', which for this single-story building are also the story drift values, are

$$\delta' = \frac{V'_A}{R_A} = \frac{52.5}{300} = 0.18 \text{ in}$$

$$\delta_B = \frac{V'_B}{R_B} = \frac{47.5}{100} = 0.48 \text{ in}$$

$$\delta_{avg} = \frac{0.18 + 0.48}{2} = 0.33 \text{ in}$$

$$\delta_{max} = \delta'_B = 0.48 \text{ in}$$

$$\frac{\delta_{max}}{\delta_{avg}} = \frac{0.49}{0.33} = 1.45 > 1.4$$

∴ Extreme Torsional Irregularity Type 1b exists. (See Example 26) Assuming SDC D, structural modeling must include 3 dimensions per §12.7.3, and diaphragm shear transfer forces to collectors must be increased 25 percent per §12.3.3.4.

Section 12.8.4.3 requires the evaluation and application of the torsional amplification factor

$$A_x = \left(\frac{\delta_{max}}{1.2_{avg}}\right)^2 = \left(\frac{0.48}{1.2(0.33)}\right)^2 = 1.47 < 3.0 \qquad (12.8\text{-}14)$$

Note: the factor A_x is not calculated iteratively (i.e., it is not recalculated with amplified torsion).

4. Torsional shears in walls A and B

To account for the effects of torsional irregularity, §12.8.4.2 requires that the accidental torsional moment, Ve_{acc}, be multiplied by the torsional amplification factor A_x.

Example 24 ■ *Horizontal Distribution of Shear*   *§12.8.4*

The most severe total shears result from the use of $V[e - A_x e_{acc}]$ for $V_{T,A}$ and $V[e + A_x e_{acc}]$ for $V_{T,B}$

$$V_{T,A} = \frac{100 \text{ kips}[(20 - 1.47 \times 4]20(300 \text{ kip/in})}{64 \times 10^4 (\text{kip/in})\text{ft}^2} = 13.24 \text{ kips}$$

$$V_{T,B} = \frac{100 \text{ kips}[(20 + 1.47 \times 4]60(100 \text{ kip/in})}{64 \times 10^4 (\text{kip/in})\text{ft}^2} = 24.3 \text{ kips}$$

5. Total shear in walls A and B

Total shear in each wall is the algebraic sum of the direct and torsional shear components

$$V_A = V_{D,A} - V_{T,A} = 75.0 - 13.2 = 61.8 \text{ kips}$$

$$V_B = V_{D,B} + V_{T,B} = 25.0 + 24.3 = 49.3 \text{ kips}$$

Commentary

Section 12.8.4.2 requires that the most severe load combination for each element shall be considered for design. This load combination involves the direct and torsional shears, and the "most severe" condition is as follows.

1. Where the torsional shear has the same sense, and is therefore added to the direct shear, the torsional shear shall be calculated using actual eccentricity plus the accidental eccentricity to give the largest additive torsional shear.

2. Where the torsional shear has the opposite sense to that of the direct shear and is to be subtracted, the torsional shear must be based on the actual eccentricity minus the accidental eccentricity to give the smallest subtractive shear.

The §12.8.4.3 requirement to multiply only the accidental torsional moment by A_x differs significantly from the 2000 IBC. It restores the requirements of the 1997 UBC and 1999 Blue Book.

Example 25
Amplification of Accidental Torsion §12.8.4.3

This example illustrates how to include the effects of accidental eccentricity in the lateral force analysis of a multi-story building. The structure is a five-story reinforced concrete building frame system. A three-dimensional rigid diaphragm model has been formulated for the evaluation of element actions and deformations due to prescribed loading conditions. Shear walls resist lateral forces in both directions.

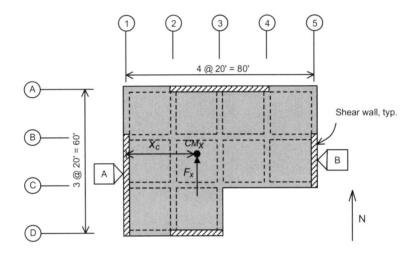

Floor plan at Level x

The lateral seismic forces F_x in the north-south direction, structure dimensions, and accidental eccentricity e_{acc} for each Level x are given below.

Level x	F_x	L_x	$\overline{x}_{CS}$	$e_{acc} = 0.05L_x$
5	110.0 kips	80.0 ft	24.2 ft	± 4.0 ft
4	82.8	80.0	25.1	± 4.0
3	65.1	80.0	27.8	± 4.0
2	42.1	80.0	30.3	± 4.0
1	23.0	80.0	31.5	± 4.0

In addition, for the given lateral seismic forces F_x a computer analysis provides the following results for the second story. Separate values are given for the application of the forces F_x at the centers of mass and the $\pm 0.05L_x$ displacements as required by §12.8.4.2. In this example, it is assumed for simplicity that the location of the center-of-mass CM_x is congruent with the center of rigidity at the level in question, resulting in zero inherent at torsion.

Example 25 ■ *Amplification of Accidental Torsion*   *§12.8.4.3*

	Force F_x Position		
	$\overline{X}_{c2}$	$\overline{X}_{c2} - e_{acc}$	$\overline{X}_{c2} + e_{acc}$
Wall shear V_A	185.0 kips	196.0 kips	174.0 kips
Wall shear V_B	115.0 kips	104.0 kips	126.0 kips
Story drift Δ_A	0.35 in	0.37 in	0.33 in
Story drift Δ_B	0.62	0.56	0.68
Level 2 displacement δ_A	0.80	0.85	0.75
Level 2 displacement δ_B	1.31	1.18	1.44

For the second story, find the following.

1. **Maximum force in shear walls A and B**

2. **Check if torsional irregularity exists**

3. **Determine the amplification factor A_x**

4. **New accidental torsion eccentricity**

Calculations and Discussion   **Code Reference**

1. **Maximum force in shear walls A and B**

The maximum force in each shear wall is a result of direct shear, inherent torsion (center of mass not being congruent with center of rigidity) and the contribution due to accidental torsion. As mentioned above, in this example it is assumed that accidental eccentricity is the only source of torsional moment at this floor level. From the above table, it is determined that

$$V_A \quad = 196.0 \text{ kips}$$

$$V_B \quad = 126.0 \text{ kips}$$

2. **Check if torsional irregularity exists**

The building may have a torsional irregularity Type 1 (Table 12.3-1). The following is a check of the story drifts.

$$\Delta_{max} \quad = 0.68 \text{ in}$$

$$\Delta_{avg} \quad = \frac{0.68 + 0.33}{2} = 0.51 \text{ in}$$

$$\frac{\Delta_{max}}{\Delta_{avg}} = \frac{0.68}{0.51} = 1.33 > 1.2$$

∴ Torsional irregularity Type 1a exists – Note: if $\Delta_{max}/\Delta_{avg}$ is larger than 1.4, then torsional irregularity Type 1b exists.

3. Determine the amplification factor A_x

Because a torsional irregularity exists, §12.8.4.3 requires that the second story torsional moment be amplified by the following factor. In this example, because the only source of torsion is the accidental eccentricity, the amplification factor will be used to calculate a new and increased accidental eccentricity, as shown below.

$$A_x = \left(\frac{\delta_{max}}{1.2\delta_{avg}}\right)^2 \qquad \text{(Eq 12.8-14)}$$

Where:

$$\delta_{max} = \delta_B = 1.44 \text{ in}$$

the average story displacement is computed as

$$\delta_{avg} = \frac{1.44 + 0.75}{2} = 1.10 \text{ in}$$

$$A_2 = \left(\frac{1.44}{(1.2)(1.10)}\right)^2 = 1.19$$

4. New accidental torsion eccentricity

Since A_2 (i.e., A_x for the second story) is greater than unity, a second analysis for torsion must be performed using the new accidental eccentricity.

$$e_{acc} = (1.19)(4.0 \text{ ft}) = 4.76 \text{ ft}$$

Example 25 ■ Amplification of Accidental Torsion §12.8.4.3

Commentary

Example calculations were given for the second story. In practice, each story requires an evaluation of the most severe element actions and a check for the torsional irregularity condition.

If torsional irregularity exists and A_x is greater than 1.0 at any level (or levels), a second torsional analysis must be performed using the new accidental eccentricities. However, it is *not* required to find the resulting new A_x values and repeat the process a second or third time (until the A_x converges to a constant or reaches the limit of 3.0). The results of the first analysis with the use of A_x are sufficient for design purposes.

While this example involves wall shear evaluation, the same procedure applies to the determination of the most severe element actions for any other lateral-force-resisting system having rigid diaphragms.

When the dynamic analysis method of §12.9 is used, all the requirements of horizontal shear distribution, given in §12.8.4, including torsion calculations that may be accounted for by displacing the calculated centers of mass of each level (§12.8.4.1 and §12.8.4.2) also apply. However, §12.9.5 states that amplification of accidental torsion, need not be amplified by A_x where accidental torsional effects are included in the dynamic analysis model. Only the accidental torsion is required to be amplified if torsional irregularity exists. Also note that A_x is not required to exceed 3.0.

Example 26
Elements Supporting Discontinuous Systems   §12.3.3.3

A reinforced concrete building has the lateral-force-resisting system shown below. Shear walls at the first-floor level are discontinuous between lines A and B and lines C and D.

The following information is given.

Seismic Design Category C
   $S_{DS} = 1.10$

Ordinary reinforced > concrete shear wall (ORCSW) building   T 12.2-1
frame system: $R = 5$ and $\Omega_o = 2.5$

Note: ORCSW not permitted in SDC D, E, or F.

Office building live load: use factor of 0.5 on L §12.4.2.3

Axial loads on column C
   D = 40 kips
   L = 20 kips
   Q_E = 100 kips

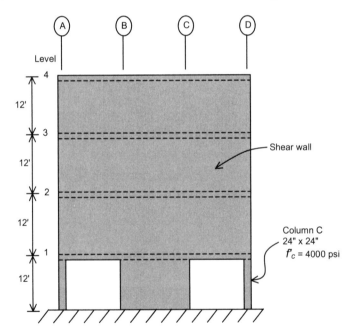

Determine the following for column C.

| 1. | Required strength |
| 2. | Detailing requirements |

Example 26 ■ *Elements Supporting Discontinuous Systems* *§12.3.3.3*

Calculations and Discussion   **Code Reference**

This example demonstrates the loading criteria and detailing required for elements supporting discontinued or offset elements of a seismic-force-resisting system.

Required strength

Because of the discontinuous configuration of the shear wall at the first story, the first story columns on lines A and D must support the wall elements above this level. Column C on line D is treated in this example. Because of symmetry, the column on line A would have identical requirements.

Section 12.3.3.3 requires that the column shall have a design strength to resist special seismic load combination of §12.4.3.2

$$P_u = 1.2D + 0.5L + 1.0E_m \qquad\qquad \text{§12.4.2.3 (Comb. 5)}$$

$$P_u = 0.9D + 1.0E_m \qquad\qquad \text{§12.4.2.3 (Comb. 7)}$$

where

$$E_m = \Omega_o \, Q_E + 0.2\, S_{DS}\, D \;= 2.5(100) + 0.2(1.10)(40) = 259 \text{ kips} \quad \text{§12.4.3.2 (Comb. 5)}$$

or $\quad E_m = \Omega_o \, Q_E - 0.2\, S_{DS}\, D \;= 2.5(100) - 0.2(1.10)(40) = 241 \text{ kips} \quad \text{§12.4.3.2 (Comb. 7)}$

Substituting the values of dead, live, and seismic loads

$$P_u \;= 1.2\,(40) + 0.5\,(20) + 259 = 317 \text{ kips compression}$$

and

$$P_u \;= 0.9\,(40) - 0.5\,(241) = -84.5 \text{ kips tension}$$

Commentary

To transfer the shears from walls A-B and C-D to the first-story wall B-C, collector beams A-B and C-D are required at Level 1. These would have to be designed according to the requirements of §12.10.2.

The load requirements of §12.3.3.3 and related sections of the relevant materials chapters apply to the following vertical irregularities and vertical elements.

1. **Discontinuous shear wall.** The wall at left has a Type 4 vertical structural irregularity. Note that only the column needs to resist the special load combinations since it supports the shear wall.

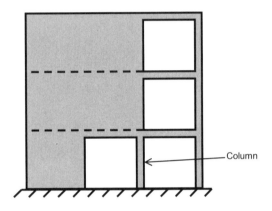

2. **Discontinuous column.** This frame has a Type 4 vertical structural irregularity.

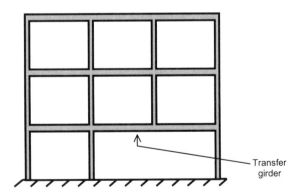

Example 26 ■ Elements Supporting Discontinuous Systems §*12.3.3.3*

3. **Out-of-plane offset.** The wall on Line A at the first story is discontinuous. This structure has a Type 4 plan structural irregularity, and §12.3.3.3 applies to the supporting columns. The portion of the diaphragm transferring shear (i.e., transfer diaphragm) to the offset wall must be designed per the requirements of §12.3.3.4. Note that the transfer diaphragm and the offset shear wall are subject to the ρ factor, but not to the special load combinations.

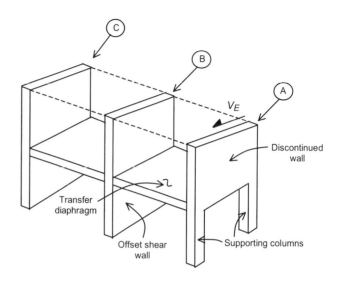

It should be noted that for any of the supporting columns shown above, the load demand E_m of §12.4.3.2 Equations 5 and 7 need not exceed the maximum force that can be transferred to the element by the lateral-force-resisting system.

Example 27
Elements Supporting Discontinuous Walls or Frames   §12.3.3.3

This example illustrates the application of the requirements of §12.3.3.3 for the allowable stress design of elements that support a discontinuous lateral-force-resisting system.

In this example, a light-framed bearing-wall building with plywood shear panels has a Type 4 vertical structural irregularity in one of its shear walls, as shown below.

The following information is given.

Seismic Design Category C

S_{DS} = 1.10
R = 6.5
Ω_o = 3.0
C_d = 4
f_1 = 0.5

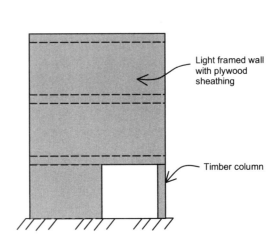

Light framed wall with plywood sheathing

Timber column

Axial loads on the timber column under the discontinuous portion of the shear wall are

Dead D = 6.0 kips
Live L = 3.0 kips
Seismic Q_E = ±7.0 kips

Determine the following.

1. **Applicable load combinations**

2. **Required column design strength**

Calculations and Discussion   **Code Reference**

1. **Applicable load combinations**

For vertical irregularity Type 4, §12.3.3.3 requires that the timber column have the design strength to resist the special seismic load combinations of §12.4.3.2. This is required for both allowable stress design and strength design. For strength design the applicable load combinations for allowable strength design are

5. $(1.2 + 0.2\,S_{DS})\,D + L + \Omega_o Q_E$

7. $(0.9 - 0.2 S_{DS})D + \Omega_o E$

Applicable load combinations for allowable strength design are:

5. $(1.0 + 0.105\,S_{DS})\,D + 0.525 + \Omega_o Q_E + 0.75L$

6. $(0.8 - 0.14\,S_{DS})\,D + 0.7\,\Omega_o Q_E$

8. $(1.0 + 0.14\,S_{DS})\,D + 0.7\,\Omega_o Q_E$

---

2. Required column design strength (strength design)

In this shear wall, the timber column carries only axial loads. The appropriate dead, live, and seismic loads are determined as

$D = 6.0$ kips

$L = 3.0$ kips

$E_m = \Omega_o\,Q_E + 0.2\,S_{DS}\,D = 3.0(7.0) + 0.2\,(1.10)\,(6.0) = 22.3$ kips

or $E_m = \Omega_o\,Q_E - 0.2\,S_{DS}\,D = 3.0(7.0) - 0.2\,(1.10)\,(6.0) = 19.7$ kips

For the required strength design-strength check, both load combinations must be checked.

$P = 1.2D + L + E_m$

$P = 1.2\,(6.0) + 0.5\,(3.0) + 22.3 = 31.0$ kips . . . (compression)

$P = 0.9D - 1.0E_m$

$P = 0.9\,(6.0) - 1.0\,(19.7) = -14.3$ kips . . . (tension)

The load factor on L in combination 5 is permitted to equal 0.5 for all occupancies in which L_o is less than or equal to 100 psf, with the exception of garages or areas occupied as places of public assembly.

Commentary

For strength design, the timber column must be checked for a compression load of 31.0 kips and a tension load of 14.3 kips.

In making an allowable stress design check, §12.4.3.3 permits use of an allowable stress increase of 1.2. The 1.2 stress increase may be combined with the duration of load increase described in the NDS. The resulting design strength = (1.2)(1.0)(1.33) (allowable stress design). This also applies to the mechanical hold-down element required to resist the tension load.

The purpose of the design-strength check is to confirm the ability of the column to carry higher and more realistic loads required by the discontinuity in the shear wall at the first floor. This is done by increasing the normal seismic load in the column Q_E by the factor $\Omega_o = 3.0$ to calculate the maximum seismic load effect E_m (§12.4.3).

Example 28 ▪ Soil Pressure At Foundation   *§§2.4; 12.13.4*

Example 28
Soil Pressure At Foundations   §§2.4; 12.13.4

Geotechnical investigation reports usually provide soil-bearing pressures on an allowable stress design basis while seismic forces in ASCE/SEI 7-05 and most concrete design (ACI/318-05, §15.2.2 and R 15.2), are on a strength design basis. The purpose of this example is to illustrate footing design in this situation.

A spread footing supports a reinforced concrete column. The soil classification at the site is sand (SW).

The following information is given.

Seismic Design Category C
> $S_{DS} = 1.0$, $I = 1.0$
> ρ   = 1.0 for structural system
> $P_D = 50$ kips

P_D includes the footing and imposed soil weight)
> $P_L = 30$ kips
> $P_E = \pm 40$ kips, $V_E = 25$ kips,

(these are the Q_E loads due to base shear V)
> Snow load $S = 0$
> Wind load $W < Q_E/1.4$

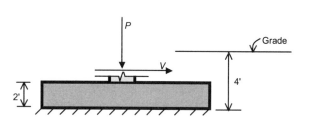

The seismic loads are from an equivalent lateral analysis.

The loads given above follow the sign convention shown in the figure.

Perform the following tasks.

1. **Determine the design criteria and allowable bearing pressure**

2. **Determine footing size**

3. **Determine soil pressure reactions for strength design of the footing section**

Calculations and Discussion   **Code Reference**

1. **Determine the design criteria and allowable bearing pressure**   **§2.4**

The seismic-force reactions on the footing are based on strength design. However, allowable stress design may be used for sizing the foundation using the load combinations given in §2.4.1.

$$D + 0.7E \qquad \qquad \text{(Comb. 5)}$$

$$D + 0.75 (0.7E + L) \qquad \qquad \text{(Comb. 6)}$$

$$0.6D + 0.7E \qquad \qquad \text{(Comb. 7)}$$

Section 12.13.4 permits reduction of overturning effects at the foundation-soil interface by 25 percent (if an equivalent lateral force analysis is used) or 10 percent (if modal analysis is used). Therefore, for the soil pressure the seismic effect is reduced

$$D + 0.75(0.7E) \qquad \qquad \text{(Comb. 5)}$$

$$D + 0.75[0.7(0.75)E + L] \qquad \qquad \text{(Comb. 6)}$$

$$0.6D + 0.7(0.75)E \qquad \qquad \text{(Comb. 7)}$$

Because foundation investigation reports for buildings typically specify bearing pressures on an allowable stress design basis, criteria for determining footing size are also on this basis.

The earthquake loads to be resisted are specified in §12.4.2 by

$$E = E_h + E_v \qquad \qquad \text{(Eq 12.4-1)}$$

Per §12.4.2.2, $E_v = 0$ for determining soil pressure. Equation 12.4-1 reduces to

$$E = E_h = \rho Q_E = (1.0)\, Q_E \qquad \qquad \text{(Eq 12.4-3)}$$

For the sand class of material and footing depth of 4 feet, the allowable gross foundation pressure p_a from a site-specific geotechnical investigation recommendation is

$$p_a = 2.40 \text{ ksf} \quad \text{for sustained loads and}$$

$$p_a = 3.20 \text{ ksf} \quad \text{for transient loads, such as seismic.}$$

Example 28 ■ *Soil Pressure At Foundation*   *§§2.4; 12.13.4*

2. Determine footing size

$$P = D + 0.75(0.7E) = 50 + 0.75(0.7)(40) = 71 \text{ kips} \qquad \text{(Comb. 5)}$$

$$P = D + 0.75[0.7(0.75)E + L] \qquad \text{(Comb. 6)}$$

$$= 50 + 0.75[0.7(0.75)40 + 30] = 88 \text{ kips}$$

$$P = 0.6D + 0.7(0.75)E \qquad \text{(Comb. 7)}$$

$$= 0.6(50) + 0.7(0.75)(-40) = 9 \text{ kips}$$

Equation 6 governs. The required footing size is 88 kips/3.20 ksf = 27.5 sf
Use 5 ft, 3-in-square footing. A = 27.6 sf

3. Determine soil pressure reactions for strength design of footing

For the design of the concrete elements, strength design is used. The reduction in overturning does not apply, and the vertical seismic load effect is included

$$P = 1.2D + 0.5L + E \qquad \text{§2.3.2 (Comb. 5)}$$

$$= 1.2(50) + 0.5(30) + 40 + 0.2(1.0)(50) = 125k$$

A uniform pressure of 125k/27.6 sf = 4.53 ksf should be used to determine the internal forces of the footing. (Note that if the footing also resisted moments, the pressure would not be uniform.)

The other seismic load combination is

$$P = 0.9D - E \qquad \text{§2.3.2 (Comb. 7)}$$

$$= 0.9(50) - 40 - 0.2(1.0)50 = -5k$$

Note that this indicates uplift will occur. ASCE/SEI 7-05 does not require that foundation stability be maintained using strength-level seismic forces. This combination is only used here to determine internal forces of concrete elements of the foundation. As it results in no internal forces, it may be neglected.

Example 29
Drift
§12.8.6

A four-story special moment-resisting frame (SMRF) building has the typical floor plan as shown below. The typical elevation of Lines A through D is also shown, and the structure does not have horizontal irregularity Types 1a or 1b.

The following information is given.

Occupancy Importance Category I

Seismic Design Category D

$$I = 1.0$$
$$C_d = 5.5$$
$$T = 0.60 \text{ sec}$$

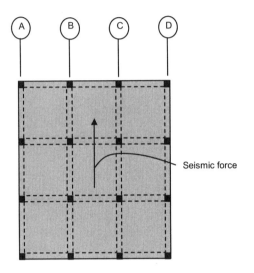

Typical floor plan

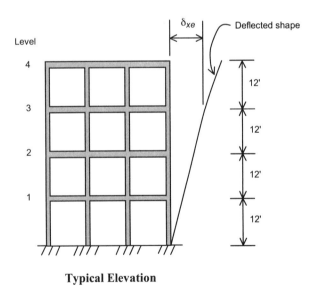

Typical Elevation

The following are the deflections (computed from static analysis – effects of *P*-delta have been checked) δ_{xe} at the center-of-mass of each floor level. These values include both translational and torsional (with accidental eccentricity) effects. As required by §12.8.6.2, δ_{xe} has been determined in accordance with design forces based on the computed fundamental period without the upper limit ($C_u T_a$) of §12.8.2.

Example 29 ■ *Drift* §12.8.6

Level	δ_{se}
4	1.51 in
3	1.03
2	.63
1	.30

For each floor-level center-of-mass, determine the following.

1. Maximum inelastic response deflection δ_{se}

2. Design story drift Δ in story 3

3. Check story 3 for story-drift limit

Calculations and Discussion **Code Reference**

1. Maximum inelastic response deflection δ_x §12.8

These are determined using the δ_{se} values and the C_d factor

$$\delta_x = \frac{C_d \delta_{xe}}{I} = \frac{5.5 \delta_{xe}}{1.0} = 5.5 \delta_{xe}$$ (Eq 12.8-15)

Therefore:

Level	δ_{xe}	δ_x
4	1.51 in	8.31 in
3	1.03	5.67
2	0.63	3.47
1	0.30	1.65

2. Design story drift Δ in story 3 due to δ_x

Story 3 is located between Levels 2 and 3.

Thus: $\Delta_3 = 5.67 - 3.47 = 2.20$ in

| 3. | **Check story 3 for story-drift limit** | **§12.12.1** |

For this four-story building with Occupancy Importance Category I, §12.12.1, Table 12.12-1 requires that the design story drift Δ shall not exceed 0.025 times the story height.

For story 3

$$\Delta_3 = 2.20 \text{ in}$$

Story-drift limit = 0.025 (144) = 3.60 in > 2.20 in

∴ Story drift is within the limit.

Example 30 ■ *Story Drift Limitations* §*12.12*

Example 30
Story Drift Limitations §12.12

For the design of new buildings, the code places limits on the design story drifts, Δ. The limits are based on the design earthquake displacement or deflection δ_x and not the elastic response deflections δ_{xe} corresponding to the design lateral forces of §12.8.

In the example given below, a four-story steel special moment-resisting frame (SMF) structure has the design force deflections δ_{xe} as shown. These have been determined according to §12.8, using a static, elastic analysis.

Occupancy Category I

Seismic Design Category D

$I = 1.0$

$C_d = 5.5$

$\rho = 1.3$

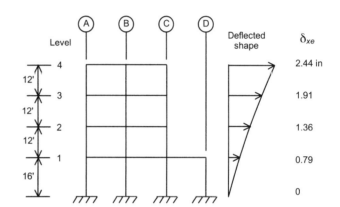

Determine the following.

1. **Design earthquake deflections δ_x**

2. **Compare design story drifts with the limit value**

Calculations and Discussion
Code Reference

1. **Design earthquake deflections δ_x** §12.8.6

The design earthquake deflections δ_x are determined from the following

$$\delta_x = \frac{C_d \delta_{xe}}{I}$$ (Eq 12.8-15)

$$\therefore \delta_x = \frac{5.5\delta_{xe}}{1.0} = 5.5\delta_{se}$$

2. Compare story drifts with the limit value   §12.12

For this four-story building in Occupancy Category I, §12.12, Table 12.12-1 requires that the calculated design story drift shall not exceed 0.025 times the story height.

For SMF in SDC D, E, and F, this limit is reduced by ρ per §12.12.1.1:

$$\Delta a/\rho = 0.025h/1.3 = 0.0192h$$

Determine drift limit at each level

Levels 4, 3, and 2

$$\Delta \leq 0.0192h = 0.0192 \, (12 \text{ ft} \times 12 \text{ in/ft}) = 2.76 \text{ in}$$

Level 1

$$\Delta \leq 0.0192h = 0.0192 \, (16 \text{ ft} \times 12 \text{ in/ft}) = 3.68 \text{ in}$$

For $\Delta = \delta_x - \delta_{x-1}$, check actual design story drifts against limits

Level x	δ_{xe}	δ_x	Δ	Limit	Status
4	2.08 in	11.43 in	2.51 in	2.76	*o.k.*
3	1.62	8.92	2.68	2.76	*o.k.*
2	1.13	6.24	2.65	2.76	*o.k.*
1	0.65	3.59	3.59	3.68	*o.k.*

Therefore: The story drift limits of §12.12 are satisfied.

Note that use of the drift limit of 0.025h requires interior and exterior wall systems to be detail to accommodate this drift per Table 12.12-1

Commentary

Whenever the dynamic analysis procedure of §12.9 is used, story drift should be determined as the modal combination of the story-drift value for each mode.
Determination of story drift from the difference of the combined mode deflections may produce erroneous results because differences in the combined modal displacements can be less than the corresponding combined modal story drift.

Example 31 ■ Vertical Seismic Load Effect   *§12.4.2.2*

Example 31
Vertical Seismic Load Effect   §12.4.2.2

Find the vertical seismic load effect, E_v, on the non-prestressed cantilever beam shown below.

The following information is given.

Seismic Design Category D

Beam unit weight = 200 plf
$S_{DS} = 1.0$

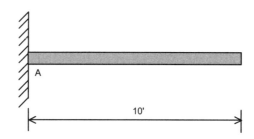

Find the following for strength design.

| **1.** | **Upward seismic forces on beam** |
| **2.** | **Beam end reactions** |

Calculations and Discussion   **Code Reference**

1. **Upward seismic forces on beam**   **§12.4.2.2**

For SDC D, the design of horizontal cantilever beams must consider

1. The governing load combination including E as defined in §12.4.2

$$E = E_h + E_v \qquad \text{(Eq 12.4-1)}$$

$$E = E_h - E_v \qquad \text{(Eq 12.4-2)}$$

$$E_h = \rho Q_E \qquad \text{(Eq 12.4-3)}$$

$$E_v = 0.2 S_{DS} D \qquad \text{(Eq 12.4-4)}$$

$Q_E = 0$ for vertical load, giving

$$E = 0 - 0.2(1.0)\, D = -0.2D$$

where the negative sign is for an upward action.

The governing load combination including the upward seismic effect from §2.3.2, (7) is

$$q_E = 0.9D + 1.0E = 0.9D + (-0.2D)$$

$$= 0.7D$$

$$= 0.7(200 \text{ plf})$$

$$= 140 \text{ plf downward}$$

$$\therefore \text{ no net upward load.}$$

The governing load combination including the downward seismic effect from §2.3.2, (5) is

$$q_E = 1.2D + 1.0E + L + 0.2S$$

$$= 1.2D + 1.0(0.2)(1.0)D + 0 + 0$$

$$= 1.4D$$

$$= 1.4(200 \text{ plf})$$

$$= 280 \text{ plf downward}$$

$$\therefore \text{ this is the maximum downward load on the beam.}$$

2. A minimum net upward seismic force. The terminology of "net upward seismic force" is intended to specify that gravity load effects cannot be considered to reduce the effects of the vertical seismic forces and that the beam must have the strength to resist the actions caused by this net upward force without consideration of any dead loads. This force is computed as 0.2 times the dead load

$$q_E = -0.2w_D = -0.2(200) = -40 \text{ plf} \qquad \text{§12.4.2.2}$$

Example 31 ■ *Vertical Seismic Load Effect*   *§12.4.2.2*

| 2. | **Beam end reactions for upward force of 40 plf** |

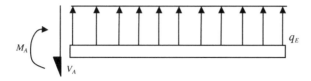

$$V_A = q_E \ell = 40 \text{ plf (10 ft)} = 400 \text{ lb}$$

$$M_A = q_E \frac{\ell^2}{2} = \frac{40(10)^2}{2} = 2000 \text{ lb/ft}$$

The beam must have strengths ϕV_n and ϕM_n to resist these actions, and the actions due to the applicable gravity load combinations.

Example 32
Design Response Spectrum   §11.4.5

Determine the general design response spectrum for a site where the following spectral response acceleration parameters have been evaluated according to the general procedure of §11.4.

$S_{DS} = 0.45g$
$S_{D1} = 0.28g$
$T_L = 8$ sec

1. Determine design response spectrum

Calculations and Discussion   *Code Reference*

Section 11.4.5 provides the equations for the 5-percent damped acceleration response spectrum S_a for the period T intervals of

$0 \leq T \leq T_o$, and $T > T_s$

T_o and T_s are calculated as

$$T_o = \frac{0.2S_{D1}}{S_{DS}} = \frac{0.2(0.28)}{0.45} = 0.12 \text{ sec}$$

$$T_s = \frac{S_{D1}}{S_{DS}} = \frac{0.28}{0.45} = 0.62 \text{ sec}$$

Example 32 ■ *Design Response Spectrum*   *§11.4.5*

The spectral accelerations are calculated as

1. For the interval $0 \leq T \leq T_o$

$$S_a = 0.6 \frac{S_{DS}}{T_o} T + 0.4 S_{DS} \qquad \text{(Eq 11.4-5)}$$

$$= 0.6 \left(\frac{0.45g}{0.12} \right) T + 0.4(0.45g)$$

$$= [2.25T + 0.18]g$$

2. For $T_o < T \leq T_s$

$$S_a = S_{DS} = 0.45g$$

3. For $T_s < T \leq T_L$

$$S_a = \frac{S_{D1}}{T} = \frac{0.28}{T} g \qquad \text{(IBC Eq 16-21)}$$

4. For $T \geq T_L$

$$S_a = S_{D1} \frac{T_L}{T^2} = \frac{2.24g}{T^2}$$

From this information the elastic design response spectrum for the site can be drawn as shown in Figure 33.1 below, per Figure 11.4-1, in ASCE/SEI 7-05

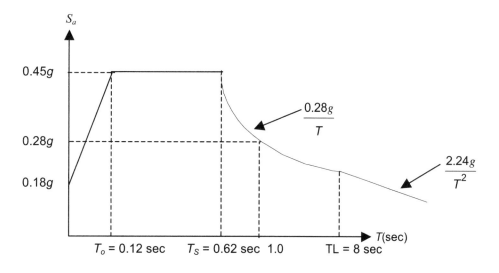

Figure 33.1 Elastic response spectrum

Example 33
Dual Systems   §12.2.5.1

This example illustrates the determination of design lateral forces for the two basic elements of a dual system. §12.2.5.1 prescribes the following features for a dual system.

1. Resistance to lateral load is provided by the combination of the moment frames and by shear walls or braced frames. Recall that the moment-resisting frames provided must be able to resist at least 25 percent of the design forces.

2. The two systems are designed to resist the total design base shear in proportion to their relative rigidities.

In present practice, the frame element design loads for a dual system are usually a result of a computer analysis of the combined frame-shear wall system.

In this example, the Equivalent Lateral-Force-Procedure of §12.8 has been used to determine the seismic demand Q_E at point A in the dual system of the building shown below. This is the beam moment M_{QE}.

The following information is given.

Seismic Design Category D
$\rho = 1.0$
$I = 1.0$

Design Base Shear
$V = 400$ kips
$Q_E = M_{QE} = 53.0$ kip-ft

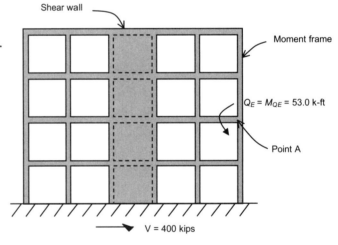

From the results of the computer analysis

ΣV shear walls	= 355 kips
ΣV columns	= 45
Total design base shear	= 400 kips

Determine the following for the moment frame system.

1. **Design criteria**

2. **Seismic design moment at A = M'_{QE}**

Example 33 ▪ *Dual Systems*　　§*12.2.5.1*

Calculations and Discussion　　　　　　　　　　　　　　　**Code Reference**

1. Design criteria

According to the two listed requirements, the moment frame must be designed for the greater value of either the Q_E value due to the design base shear V loading on the combined frame – shear wall system, or the Q'_E value resulting from at least 25 percent of the design forces. This 25-percent requirement may be interpreted in two ways.

a. Q'_E may be found by an equivalent lateral-force analysis of the independent moment frame using 25 percent of the design base shear V.

b. Q'_E may be found by factoring the combined frame-shear wall system Q'_E value such that Q'_E corresponds to the action that would occur if the portion of the base shear resisted by the moment frame V_F were to be at least equal to 25 percent of the design base shear V.

2. Seismic Design Moment at A = M'_{QE}

It is elected to use the factored Q_E (option b) listed above, because this procedure includes the interaction effects between the frame and the shear wall

From the combined frame-shear wall analysis with forces due to the design base shear $V = 400$ kips, the portion V_F of the base shear resisted by the moment-frame is equal to the sum of the first story frame column shears in the direction of loading. For this example, assume that

$$V_F = \Sigma V_{col} = 45 \text{ kips}$$

The required values Q'_E corresponding to a frame base shear resistance equal to 25 percent of V is given by

$$Q'_E = \frac{0.25V}{V_F}(Q_E)$$

and the seismic design moment at A is

$$M'_{QE} = \frac{0.25(400)}{45}(53.0) = 117.8 \text{ kip-ft}$$

Commentary

Use of a dual system has the advantage of providing the structure with an independent vertical load-carrying system capable of resisting 25 percent of the design base shear, while at the same time the primary system, either shear wall or braced frame, carries its proportional share of the design base shear. For this configuration, the code permits use of a larger R value for the primary system than would be permitted without the 25-percent frame system.

Design Criterion 1a involving the design of the moment frame independent from the shear wall or bracing system for 25 percent of the design base shear should be considered for high-rise buildings. The slender configuration of the shear walls or bracing systems can actually load the moment frame at the upper levels of the combined model, and excessively large moment frame design actions would result from the use of Design Criterion 1b, where these large actions would be multiplied by $\dfrac{0.25V}{V_F}$

Example 34 ■ *Lateral Forces for One-Story Wall Panels* §12.11

Example 34
Lateral Forces for One-Story Wall Panels §12.11

This example illustrates the determination of the total design lateral seismic force on a tilt-up wall panel supported at its base and at the roof diaphragm level. Note that the panel is a bearing wall and shear wall.

For the tilt-up wall panel shown below, determine the out-of-plane seismic forces required for the design of the wall section. This is usually done for a representative 1-foot width of the wall length, assuming a uniformly distributed out-of-plane loading.

The following information is given.

Seismic Design Category D

$$I = 1.0$$
$$S_{DS} = 1.0g$$

Panel thickness = 8 inches
Normal weight concrete (150 pcf)

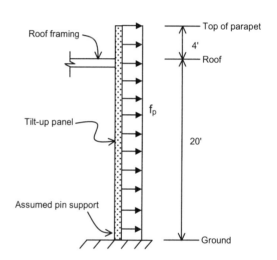

Determine the following.

1.	Out-of-plane force for wall panel design
2.	Shear and moment diagrams for wall panel design
3.	Loading, shear and moment diagrams for parapet design

Calculations and Discussion **Code Reference**

1. **Out-of-plane force for wall panel design** **§12.11**

Under §12.11.1, the design lateral loading is determined using

$$F_p = 0.40 \, S_{DS} I \, w_p \geq 0.1 \, w_p$$

where w_p is the weight of the wall.

Per §12.11.2, the force must be taken as no less than 400 lb/ft $S_{DS}I$, nor less than 280 lb/ft

Note that if the diaphragm is flexible, §12.11.2.1 requires the anchorage force (but not the wall force) to be increased.

The force F_p is considered to be applied at the mid-height (centroid) of the panel, but this must be uniformly distributed between the base and the top of parapet.

For the given $S_{DS} = 1.0$ and $I = 1.0$, the wall panel seismic force is

$$F_p = 0.40(1.0)(1.0)w = 0.40w$$

The weight of the panel between the base and the top of the parapet is

$$w_w = \left(\frac{8}{12}\right)(150)(24) = 2400 \text{ lb per foot of width}$$

$$F_p = 0.40\,(2400) = 960 \text{ lb/ft}$$

$$F_p > 400 \text{ lb/ft } S_{DS}I = 400(1)(1) = 400 \text{ lb/ft}$$

$$F_p > 280 \text{ lb/ft}$$

The force F_p is the total force on the panel. It acts at the centroid. For design of the panel for out-of-plane forces, F_p must be expressed as a distributed load f_p

$$f_p = \frac{960 \text{ lb/ft}}{24 \text{ ft}} = 40.0 \text{ plf/ft}$$

2. **Shear and moment diagrams for wall panel design**

Using the uniformly distributed load f_p, the loading, shear, and moment diagrams are determined for a unit width of panel. The 40.0 plf/ft uniform loading is also applied to the parapet. See step 3, below, for the parapet design load.

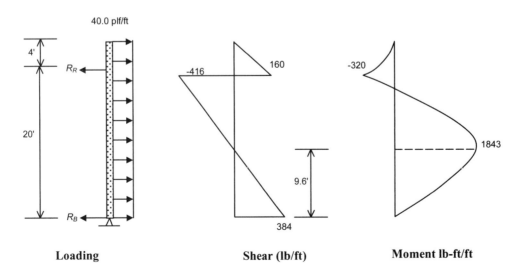

| Loading | Shear (lb/ft) | Moment lb-ft/ft |

Example 34 ■ Lateral Forces for One-Story Wall Panels §12.11

When the uniform load is also applied to the parapet, the total force on the panel is

$$40.0 \text{ plf/ft } (24 \text{ ft}) = 960 \frac{\text{lb}}{\text{ft}}$$

The reaction at the roof and base are calculated as

$$R_R = \frac{960(12)}{20} = 576 \text{ lb/ft}$$

$$R_B = 960 - 576 = 385 \text{ lb/ft}$$

The shears and moments are the Q_E load actions for strength design. Note that the reaction at the roof R_R is not necessarily the force used for wall-to-roof anchorage design, see 12.11.

---

3. Loading, shear and moment diagrams for parapet design §13.3.1

This section requires that the design force for parapets (note that parapets are classified as architectural components) be determined by Equation 13.3-1 with the Table 13.5-1 values of

$$a_p = 2.5 \text{ and } R_p = 2.5 \qquad \text{T 13.5-1}$$

for the unbraced cantilever parapet portion of the wall panel.

The parapet is considered an element with an attachment elevation at the roof level

$$z = h$$

The weight of the parapet is

$$W_p = \left(\frac{8}{12}\right)(150)(4) = 400 \text{ lb per foot of width}$$

The concentrated force applied at the mid-height (centroid) of the parapet is

$$F_p = \frac{0.4 a_p S_{DS} I_p}{R_p}\left(1 + 2\frac{z}{h}\right)W_p \qquad \text{(Eq 13.3-1)}$$

$$F_p = \frac{0.4(2.5)(1.0)(1.0)}{2.5}\left(1 + 2\frac{20}{20}\right)W_p$$

$$F_p = 1.2 W_p = 1.2\,(400) = 480 \text{ lb/ft} < 1.6\, S_{DS} I_p W_p = 640 \text{ lb/ft} \ldots o.k. \qquad \text{(Eq 13.3-2)}$$

and $> 0.3\, S_{DS} I_p W_p \ldots o.k.$ (Eq 13.3-3)

The equivalent uniform seismic force is

$$f_p = \frac{480}{4} = 120 \text{ plf/ft for parapet design}$$

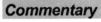

120 plf/ft

4'

R_R

Loading

480

Shear (lb/ft)

-960

Moment (lb-ft/ft)

Commentary

Note that for a large portion of the lower south-east region of the USA (Texas, Arkansas, Louisiana, Mississippi, Alabama, Georgia, and Florida) the minimum wind forces may govern over the seismic forces.

Example 35 Out-of-Plane Seismic Forces for Two-Story Wall Panel   §12.11.1 and 12.11.2

This example illustrates determination of out-of-plane seismic forces for the design of the two-story tilt-up wall panel shown below. A typical solid panel (no door or window openings) is assumed. Walls span from floor to floor to roof. The typical wall panel in this building has no pilasters and the tilt-up walls are bearing walls. The roof consists of 1-1/2-inch, 20-gage metal decking on open web steel joists and has been determined to be a flexible diaphragm. The second floor consists of 1-inch, 18-gage composite decking with a 2-1/2-inch lightweight concrete topping. This is considered a rigid diaphragm.

The following information is given.

Seismic Design Category D

$$S_{DS} = 1.0$$
$$I = 1.0$$

Wall weight $= W_W = 113$ psf

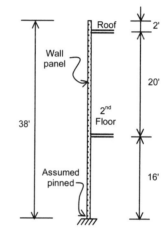

Wall section

Determine the following.

1. **Out-of-plane forces for wall panel design**

2. **Out-of-plane forces for wall anchorage design**

1. Out-of-plane forces for wall panel design   §12.11.1

Requirements for out-of-plane seismic forces are specified in §12.11.1

$$F_p = 0.40\,S_{DS}Iw_w \geq 0.1w_w$$

$$= 0.40(1.0)(1.0)w_w = 0.40w_w = 0.40(113)$$

$$= 45.2 \text{ psf}$$

For a representative 1-foot-wide strip of wall length, F_p is applied as a uniform load

$$f_p = F_p(1 \text{ ft}) = 45.2 \text{ plf}$$

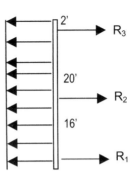

For the purpose of wall design, the required shears and moments may be evaluated by using reaction values based on the tributary area for the 1-ft strip

$$R_1 \left(\frac{16 \text{ ft}}{2}\right) f_p = 8(45.2) = 362 \text{ lb}$$

$$R_2 \left[\left(\frac{16}{2}\right) + \left(\frac{20}{2}\right)\right] f_p = 18(45.2) = 814 \text{ lb}$$

$$R_3 = \left[\left(\frac{20}{2}\right) + 2\right] f_p = 12(45.2) = 542 \text{ lb}$$

Note that the 2-foot-high parapet must be designed for the seismic force F_p specified in §13.3.1, with height z at parapet centroid 37 ft, $a_p = 2.5$ and $R_p = 2.5$

---

2. | Out-of-plane forces for wall anchorage design   §12.11.2.1

a. | Anchorage force for the flexible roof diaphragm

$$F_p = 0.8 \, S_{DS} I w_w \tag{Eq 12.11-1}$$

where w_w is the weight of the wall tributary to the anchor

$$w_w = \left[\left(\frac{20 \text{ ft}}{2}\right) + 2 \text{ ft}\right](113 \text{ psf}) = 1356 \text{ plf}$$

$$F_p = 0.8(1.0)(1.0)(1356) = 1085 \text{ plf}$$

The design force per anchor is F_p times the anchor spacing. For example if the spacing is at 4 feet, the anchor must be designed for $(1085)(4 \text{ ft}) = 4340 \text{ lb}$.

b. Anchorage force for the rigid second floor diaphragm   §12.11.2

For the case of rigid diaphragms the anchorage force is given by the greater of the following:

a. The force set forth in §12.11.1. $F_p = 0.4\ S_{DS}I$

b. A force of 400 $S_{DS}I$ (plf).

c. 280 (plf) of wall.

> z = 16 ft = the height of the anchorage of the rigid diaphragm attachment, and W_p is the weight of the wall tributary to the anchor

$$W_p = \left[\left(\frac{20\ ft}{2}\right) + \left(\frac{16\ ft}{2}\right)\right](113\ psf) = 2034\ plf$$

$$F_p = 0.4(1.0)(1.0) = W_p = 0.4(2034)$$
$$= 814\ plf$$

$$F_p = 400\ S_{DS}I = 400(1.0)(1.0)$$
$$= 400\ plf$$

$$F_p = 280\ plf$$

$$\therefore F_p = 814\ plf\ controls$$

Commentary

For flexible or rigid diaphragms for all seismic design categories (SDCs), the seismic out-of-plane forces for the design of the wall are not dependent on the height of the wall in relationship to the total height of the building, §12.11.

For flexible diaphragms of SDCs A and B, the seismic anchorage forces are given in §12.11.2 and for SDCs C, D, E, and F, the seismic anchorage forces are given in §12.11.2.1.

For rigid diaphragms of SDCs A and B, the seismic anchorage forces are given in §12.11.2.

For rigid diaphragms of SDCs C, D, E, and F, the seismic anchorage forces are given in §12.11.2.

Example 36 ■ *Rigid Equipment*   *§13.3.1*

Example 36
Rigid Equipment   §13.3.1

This example illustrates determination of the design seismic force for the attachments of rigid equipment (see commentary). Attachment, as used in the code, means those components, including anchorage, bracing, and support mountings, that "attach" the equipment to the structure.

The three-story building structure shown below has rigid electrical equipment supported on nonductile porcelain insulators that provide anchorage to the structure. Identical equipment is located at the base and at the roof of the building.

Seismic Design Category D

S_{DS} = 1.1
I_p = 1.0
W_p = 10 kips

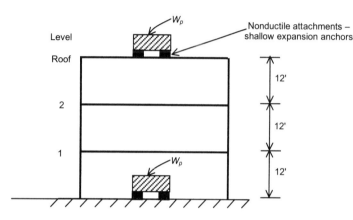

Find the following.

1. **Design criteria**

2. **Design lateral seismic force at base**

3. **Design lateral seismic force at roof**

Calculations and Discussion   **Code Reference**

1. **Design criteria**   §13.3-1

The total design lateral seismic component force to be transferred to the structure is determined from

$$F_P = \frac{0.4 a_p S_{DS} W_p}{(R_p / I_p)}\left[1 + 2\left(\frac{z}{h}\right)\right]$$ (Eq 13.3-1)

Values of a_p and R_p are given in Table 13.6-1. Also note that for shallow expansion anchors $R_p = 1.5$, see §13.4.2.

$$a_p = 1.0, R_p = 1.5 \qquad \text{T 13.6-1}$$

2. Design lateral seismic force at base §13.3-1

$$z_x = 0$$

$$F_p = \frac{0.4(1.0)(1.1)(10 \text{ kips})}{(1.5/1.0)}\left[1 + 2\left(\frac{0}{36}\right)\right] = 2.93 \text{ kips}$$

Also §13.3.1 has a requirement that F_p be not less than $0.3\,S_{DS}\,I_p W_p$

Check $F_p \leq 0.3\,S_{DS}\,I_p W_p = 0.3\,(1.1)\,(1.0)\,10 = 3.3 \text{ kips}$ (Eq 13.3-3)

$\therefore F_p = 3.3 \text{ kips} \dots$ Equation 13.3-3 governs

3. Design lateral seismic force at roof

$$z_x = h_r = 36 \text{ ft}$$

$$F_p = \frac{0.4(1.0)(1.1)(10 \text{ kips})}{(1.5/1.0)}\left[1 + 2\left(\frac{36}{36}\right)\right] = 8.8 \text{ kips}$$

Section 13.3.1 states that F_p need not exceed $1.6\,S_{DS}\,I_p\,W_p$

Check $F_p \leq 1.6\,S_{DS}\,I_p\,W_p = 1.6\,(1.1)\,(1.0)\,10 = 17.6 \text{ kips}$ (Eq 13.3-2)

$\therefore F_p = 8.8 \text{ kips} \dots$ Equation 16-67 governs.

Commentary

The definition of a rigid component (e.g., item of equipment) is given in §11.2. Rigid equipment (including its attachments; anchorages, bracing, and support mountings) that has a period less than or equal to 0.06 seconds.

Example 36 ■ *Rigid Equipment* §13.3.1

The fundamental period T_p for mechanical and electrical equipment shall be determined by the formula given in §13.6.2

$$T_p = 2\pi \sqrt{\frac{W_p}{K_p g}}$$ (Eq 13.6-1)

Where:

g = acceleration of gravity in inches/sec^2
K_p = stiffness of resilient support system
T_p = component fundamental period
W_p = component operating weight

The component anchorage design force F_p (i.e., the force in the connected part) is a function of $1/R_p$, where $R_p = 1.5$ for shallow anchors, (see §13.4.2).

Generally, only equipment such as anchorage or attachments or components need be designed for seismic forces. This is discussed in §13.1.4. Where equipment, which can be either flexible or rigid, comes mounted on a supporting frame that is part of the manufactured unit, the supporting frame must also meet the seismic design requirements of §13.

Note that §13.2.5 allows testing as an alternative to the analytical methods of §13. Testing should comply with ICC-ES AC156.

Section 13.1.3 requires a component importance factor greater than 1.0 ($I_p = 1.5$) for the following.

- Life safety component required to function after an earthquake
- Components of hazardous materials
- Occupancy Category IV components needed for continued operation of the facility

Example 37
Flexible Equipment   §13.3.1

This example illustrates determination of the design seismic force for the attachments of flexible equipment, see commentary. Attachment as used in the code means those components, including anchorage, bracing, and support mountings, that "attach" the equipment to the structure.

The three-story building structure shown below has flexible air-handling equipment supported by a ductile anchorage system. Anchor bolts in the floor slab meet the embedment length requirements. Identical equipment is located at the base and at the roof of the building.

Seismic Design Category D

$S_{DS} = 1.1$
$I_p = 1.0$
$W_p = 10$ kips

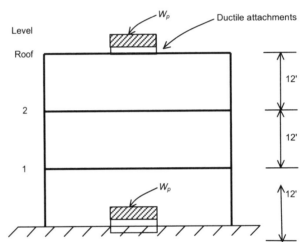

Find the following.

1.	Design criteria
2.	Design lateral seismic force at base
3.	Design lateral seismic force at roof

Calculations and Discussion   **Code Reference**

1. Design criteria   §13.3.1

The total design lateral seismic component force to be transferred to the structure is determined from

$$F_p = \frac{0.4a_p S_{DS} W_p}{R_p / I_p}\left[1 + 2\frac{z}{h}\right]$$ (Eq 13.3-1)

Example 37 ■ *Flexible Equipment* §*13.3.1*

Values of a_p and R_p are given in Table 13.6-1. Since the equipment is flexible and has limited deformability elements and attachments

$$a_p = 2.5, R_p = 2.5 \qquad \text{T 13.6-1}$$

---

2. **Design lateral seismic force at base**

$$z = 0$$

$$F_p = \frac{0.4(2.5)(1.1)(10 \text{ kips})}{(2.5/1.0)}\left[1 + 2\frac{0}{36}\right] = 4.4 \text{ kips}$$

Section 13.3.1 has a requirement that F_p be not less than $0.3\,S_{DS}I_pW_p$ (Eq 13.3-3)

Check $F_p \geq 0.3\,S_{DS}I_pW_p = 0.3(1.1)(1.0)(10) = 3.3 \text{ kips}$

$\therefore F_p = 4.4 \text{ kips…Eq 13.3-1 governs.}$

---

3. **Design lateral seismic force at roof**

$$z = h = 36 \text{ ft}$$

$$F_p = \frac{0.4(2.5)(1.1)(10 \text{ kips})}{(2.5/1.0)}\left[1 + 2\frac{36}{36}\right] = 13.2 \text{ kips}$$

Section 13.3.1 states that F_p need not exceed $1.6\,S_{DS}\,I_pW_p$ (Eq 13.3-2)

Check $F_p \leq 1.6\,S_{DS}W_p = 1.6(1.1)(10) = 17.6 \text{ kips}$

$\therefore F_p = 13.2 \text{ kips …Eq 13.3-1 governs.}$

Commentary

The definition of flexible equipment is given in §11.2. Flexible equipment (including its attachments anchorages, bracing, and support mountings), has a period greater than 0.06 second.

It should be noted that the component anchorage design force, F_p (i.e., the force in the connected part), is a function of $1/R_p$, where anchorage of any kind is shallow (see §13.4.2).

Generally, only equipment anchorage or components need be designed for seismic forces. Where the equipment, which can be either flexible or rigid, comes mounted on a supporting frame that is part of the manufactured unit, then the supporting frame must also meet the seismic design requirements of §13.3.

Also note that §13.2.1 requires that, "Architectural, mechanical, and electrical components supports and attachments shall comply with the sections referenced in Table 13.2-1."

Those architectural, mechanical, and electrical systems and their components that are part of a designated seismic system, as defined in §13.2.1, shall be qualified by either test or calculation. A certificate of compliance shall be submitted to both the registered design professional in responsible charge of the design of the designated seismic system and the building official for review and approval. ICC ES has published Acceptance Criteria (AC 156) that addresses the qualification test to satisfy the referenced code requirements.

A component importance factor greater than 1.0 ($I_p = 1.5$) is required for the following.

- Life safety component required to function after an earthquake
- Components of hazardous materials
- Occupancy Category IV components needed for continued operation of the facility

Example 38 ■ *Relative Motion of Equipment Attachments* §13.3.2

Example 38
Relative Motion of Equipment Attachments §13.3.2

Section 13.3.2 requires that the design of equipment attachments in buildings have the effects of the relative displacement of attachment points considered in the lateral force design. This example illustrates application of this requirement.

A unique control panel frame is attached to the floor framing at Levels 2 and 3 of the special steel moment frame building shown below.

The following information is given.

Seismic Design Category D
Occupancy Category II,

$$\delta_{xAe} = 1.08 \text{ in}$$
$$\delta_{yAe} = 0.72 \text{ in}$$
$$R = 8.0$$
$$C_d = 5.5$$
$$\Delta_{aA} = 0.015 h_x$$

Panel frame: $EI = 10 \times 10^4 \text{ kip-in}^2$

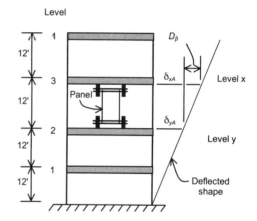

Determine the following:

1.	Story drift to be considered
2.	Induced moment and shear in frame

Calculations and Discussion Code Reference

1. **Story drift to be considered**

Section 13.3.2 requires that equipment attachments be designed for effects induced by D_p (relative seismic displacement). This is determined as follows.

$$D_p = \delta_{xA} - \delta_{yA} = 1.98 \text{ in} \qquad \text{Governs} \qquad \text{(Eq 13.3-5)}$$

where

$$\delta_{xA} = \delta_{xAe} \, C_d = 5.94 \text{ in}$$
$$\delta_{xAe} = 1.08 \text{ in}$$
$$C_d = 5.50$$

and　$\delta_{yA} = \delta_{yAe}C_d = 3.96$ in

$\delta_{yAe} = 0.72$ in
$C_d = 5.50$

Note that D_p is not required to be taken as greater than

$$D_p = (x - y)\frac{\Delta_{aA}}{h_{sx}} = (432 - 288)\frac{6.48}{432} = 2.26 \qquad \text{(Eq 13.3-6)}$$

where

x　　$= 36 \text{ ft} \times 12 = 432$ in
y　　$= 24 \text{ ft} \times 12 = 288$ in
Δ_{aA}　$= 0.015\, h_{sx} = 0.015\,(432) = 6.48$ in
h_{sx}　$= 36 \text{ ft} \times 12 = 432$ in

Thus:　$D_p = 1.98$ in

| **2.** | **Induced moment and shear in frame** | **§13.3.2** |

A liberal estimate of the moment and shear can be made using the following equations.

$$M = \frac{6EID_p}{H^2} = \frac{6(10 \times 10^4)(1.98)}{(144)^2} = 57.92 \text{ kip-in}$$

$$V = \frac{2M}{H} = \frac{57.29}{72} = 0.795 \text{ kips}$$

$$M = \frac{6\,EID_p}{H^2} \qquad V = \frac{2M}{H}$$

Commentary

The attachment details, including the body and anchorage of connectors, should follow the applicable requirements of §13.4. For example, if the anchorage is provided by shallow anchor bolts, then $R_p = 1.5$.

When anchorage is constructed of nonductile materials, $R_p = 1.0$. One example of a nonductile anchorage is the use of adhesive. Adhesive is a "glued" attachment (e.g., attachment of pedestal legs for a raised computer floor). It should be noted that attachment by adhesive is not the same as anchor bolts set in a drilled hole with an epoxy type adhesive.

Example 39   Deformation Compatibility for
Seismic Design Categories D, E, and F   §12.12.4

A two-level concrete parking structure has the space frame shown below. The designated lateral-force-resisting system consists of a two-bay special reinforced concrete moment-frame (SRCMF) located on each side of the structure. The second-level gravity load-bearing system is a post-tensioned flat plate slab supported on ordinary reinforced concrete columns,

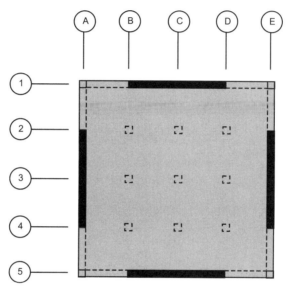

Plan at second level

The following information is given.

Seismic Design Category D

$\delta_{xe} = 0.57$ in
$R = 8.0$
$C_d = 5.5$
Column section = 12 in × 12 in
Column clear height = 12 ft
Concrete $E_c = 3 \times 10^3$ ksi
$I = 1.25$

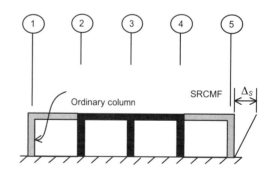

Ordinary column

SRCMF Δ_S

Elevation Line E

Find the following.

1.   **Moment in ordinary column**

2.   **Detailing requirements for ordinary column**

Calculations and Discussion   **Code Reference**

1. Moment in ordinary column   §12.12.4

Section 21.11 of ACI 318-05 specifies requirements for frame members that are not part of the designated lateral force-resisting system. The ordinary columns located in the perimeter frames, and the interior flat plate/column system, fall under these requirements and must be checked for the moments induced by the maximum inelastic response displacement. For this example, the columns on line E will be evaluated.

$$\delta_x = \frac{C_d \delta_{xe}}{I} = \frac{5.5(0.57)}{1.25} = 2.51 \text{ in}$$ (Eq 12.8-15)

The moment induced in the ordinary column due to the maximum inelastic response displacement δ_x on line E must be determined.

For purposes of this example, a fixed-fixed condition is used for simplicity. In actual applications, column moment is usually determined from a frame analysis.

$$M_{col} = \frac{6 E_c I_c \delta_x}{h^2}$$

$$h = 12 \text{ ft} \times 12 \text{ inches} = 144 \text{ in}$$

$$I_g = \frac{bd^3}{12} = 12\frac{(12)^3}{12} = 1728 \text{ in}^4$$

The cracked section moment of inertia I_c can be approximated as 50 percent of the gross section I_g. Section 21.11 of ACI 318-05 implies that the stiffness of elements that are part of the lateral-force-resisting system shall be reduced – a common approach is to use one half of the gross section properties. This requirement also applies to elements that are not part of the lateral-force-resisting system.

$$I_c = \frac{I_g}{2} = 864 \text{ in}^4$$

$$M_{col} = \frac{6(3 \times 10^3)(864)(2.51)}{(144)^2} = 1883 \text{ kip-in}$$

2. Detailing requirements for ordinary column.

Section 21.11.1 of ACI 318-05 requires that frame members, such as the column, that are assumed not to be part of the lateral-force-resisting system must be detailed according to ACI §21.11.2 or §21.11.3, depending on the magnitude of the moments induced by δ_x.

Commentary

In actual applications, the flat plate slab must be checked for flexure and punching shear due to gravity loads and the frame analysis actions induced by δ_x.

Note that this example problem shows only one way to configure this structure – that is to combine a ductile SRCMRF with an ordinary, or non-ductile, interior column. ACI requirements for this configuration stress that the non-ductile interior column must resist the structure lateral deformation by strength alone.

However, the code also permits an alternative way to configure this structure – by combining the ductile SRCMRF with <u>ductile</u> interior columns. In this configuration, if interior concrete columns are detailed according to the requirements of ACI 318 §21.11.3, then design moments resulting from lateral structure seismic displacements need not be calculated for that column at all.

Example 40
Adjoining Rigid Elements   §12.7.4

The concrete special reinforced concrete moment-resisting frame (SRCMF) shown below is restrained by the partial height infill wall that is not considered to be a part of the seismic force-resisting system. The infill is solid masonry and has no provision for an expansion joint at the column faces. The design story drift Δ was computed according to the procedure given in §12.8.6.

Seismic Design Category D

$$\Delta = 2.5 \text{ in}$$

Column properties

$$f_c' = 3000 \text{ psi}$$
$$E_c = 3 \times 10^3 \text{ ksi}$$
$$A_c = 144 \text{ in}^4$$
$$I_c = 854 \text{ in}^4$$

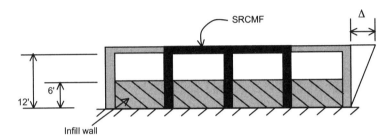

Typical elevation

Determine the following.

1. **Deformation compatibility criteria**

2. **Approximate column shear**

Calculations and Discussion   **Code Reference**

1. **Deformation compatibility criteria**   §12.7.4

The infill wall, which is not required by the design to be part of the lateral-force-resisting system, is an adjoining rigid element. Under §12.7.4, it must be shown that the adjoining rigid element, in this case the masonry infill wall, must not impair the vertical- or lateral-load-resisting ability of the SRCMF columns. Thus, the columns must be checked for ability to withstand the Δ displacement of 2.5 inches while being simultaneously restrained by the 6-foot-high infill walls.

Example 40 ■ *Adjoining Rigid Elements* **§12.7.4**

2. Approximate column shear

Column shear will be determined from the frame design story drift, Δ. For purposes of the example, the expression for shear due to a fixed-fixed condition will be used for simplicity. Also note the restrained column height is 6 ft or 72 inches.

$$V_{col} = \frac{12 E_c I_c \Delta}{h^3} = \frac{12(3 \times 10^3)(854)(2.5)}{(72)^3} = 205.9 \text{ kips}$$

Column clear height = 72 in

Because the SRCMF is the primary lateral-force-resisting system, Δ has been determined by neglecting the stiffness of the rigid masonry.

The induced column shear stress is $\frac{V_{col}}{A_c} = 1447$ psi. This is approximately $26\sqrt{f'_c}$ and would result in column shear failure. Therefore, a gap must be provided between the column faces and the infill walls. Alternately, it would be necessary to either design the column for the induced shears and moments caused by the infill wall, or demonstrate that the wall will fail before the column is damaged. Generally, it is far easier (and more reliable) to provide a gap sufficiently wide to accommodate Δ.

For this example, with the restraining wall height equal to one half the column height, the gap should be greater than or equal to $\Delta/2 = 1.25$ in. If this were provided, the column clear height would be 144 inches, with resulting column shear

$$V'_{col} = \frac{12(3 \times 10^3)(854)(2.5)}{(144)^3} = 25.7 \text{ kips}$$. This is one-eighth of the restrained column

shear of 205 kips, and corresponds to a column shear stress of approximately $3.3\sqrt{f'_c}$.

Commentary

It is also possible that the restraint of the infill walls could cause an irregularity, such as a building torsional irregularity. This should be evaluated if such restraints are present.

Example 41
Exterior Elements: Wall Panel   §13.5.3

This example illustrates the determination of the design lateral seismic force F_p on an exterior element of a building, in this case an exterior wall panel.

A five-story moment frame building is shown below. The cladding on the exterior of the building consists of precast reinforced concrete wall panels.

The following information is given.

Seismic Design Category D

$$I = 1.0$$
$$S_{DS} = 1.0$$

Panel size : 11 ft 11 in by 19 ft 11 in
Panel thickness: 6 in
Panel weight: $W_p = 14.4$ kips

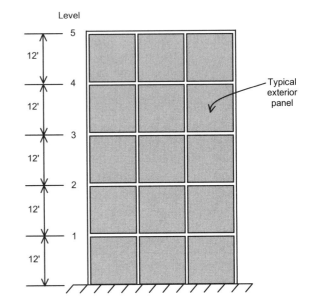

Find the following.

1.	Design criteria
2.	Design lateral seismic force on a panel at the fourth story
3.	Design lateral seismic force on a panel at the first story

---

Calculations and Discussion   Code Reference

| 1. | Design criteria | §13.5.3

For design of exterior elements, such as the wall panels on a building, that are attached to the building at two levels, design lateral seismic forces are determined from Equation 13.3-4. The panels are attached at the two elevations z_L and z_u. The intent of the code is to provide a value of F_p that represents the average of the acceleration inputs from the two attachment locations. This can be taken as the average of the two F_p values at z equal to z_L and z_u.

Example 41 ■ *Exterior Elements: Wall Panel*   §13.5.3

$$F_p = \frac{0.4 a_p S_{DS} I_p}{R_p}\left[1 + 2\frac{z}{h}\right] W_p > 0.3\, S_{DS} I_p W \qquad \text{(Eq 13.3-1)}$$

$$a_p = 1.0,\ R_p = 2.5 \qquad \text{T 13.5-1}$$

2. Design lateral seismic force on a panel at the fourth story

Assuming connections are 1 foot above and below the nominal 12-foot panel height

$$z_u = 47\ \text{ft}$$

$$z_L = 37\ \text{ft}$$

$$h = \text{ft}$$

$$F_{pU} = \frac{0.4(1.0)(1.0)(1.0)}{2.5}\left[1 + 2\left(\frac{47}{60}\right)\right] W_p = 0.411 W_p$$

$$F_{pL} = \frac{0.4(1.0)(1.0)(1.0)}{2.5}\left[1 + 2\left(\frac{37}{60}\right)\right] W_p = 0.357 W_p$$

$$F_{p4} = \frac{F_{pU} + F_{pL}}{2} = \frac{(0.411 + 0.357)}{2} W_p$$

$$F_{p4} = 0.384 W_p = (0.384)(14.4) = 5.53\ \text{kips}$$

Check: $F_{p4} > 0.3\, S_{DS} I_p W_p = 0.3(1.0)(1.0) W_p = 0.3 W_p \ldots o.k.$ (Eq 13.3-3)

Check: $F_{p4} \le 1.6\, S_{DS} I_p W_p = 1.6(1.0)(1.0) W_p = 1.6 W_p \ldots o.k.$ (Eq 13.3-2)

3. Design lateral seismic force on a panel at the first story

The following are known.

$$z_u = 11\ \text{ft}$$

$$z_L = 0$$

$$h = 60\ \text{ft}$$

$$\frac{F_{pU}}{2.5} = \frac{0.4(1.0)(1.0)(1.0)}{2.5}\left[1 + 2\left(\frac{11}{60}\right)\right]W_p = 0.219W_p$$

Check that F_{pU} is greater than $0.3\,S_{DS}I_pW_p$

$$F_{pU} = 0.3(1.0)(1.0)W_p = 0.30W_p \ldots \textbf{not } o.k.$$

Also $F_{pL} < F_{pU} < 0.30W_p$

$\therefore$ use $F_{pL} = F_{pU} = 0.30W_p$

$$F_{p1} = \frac{F_{pU} + F_{pL}}{2} = 0.30W_p = (0.30)(14.4) = 4.32 \text{ kips}$$

Commentary

Note that the design of the panel may be controlled by non-seismic load conditions of the fabrication process, transportation, and installation. Also note that the forces induced by displacement D_p from Equation 13.3-5 need to be checked per §13.3.2.1.

Example 42
Exterior Nonstructural Wall Elements: Precast Panel   §13.5.3

This example illustrates the determination of the total design seismic lateral force for the design of the connections of an exterior wall panel to a building.

An exterior nonbearing panel is located at the fourth story of a five-story moment frame building. The panel support system is shown below, where the pair of upper brackets must provide resistance to out-of-plane wind and seismic forces and in-plane vertical and horizontal forces. The panel is supported vertically from these brackets. The lower pair of rod connections provides resistance to only the out-of-plane forces.

Seismic Design Category D

$S_{DS} = 1.0$
$I_p = 1.0$
$f_1 = 0.5$
Height to roof, $h = 60$ ft
Panel weight = 14.4 kips
$\rho = 1.0$ per §12.3.4.1(3).
Panel live load, $L = 0$

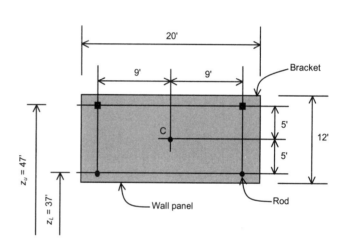

Find the following.

1. Strength design load combinations

2. Lateral seismic force at center-of-mass C of panel

3. Combined dead and seismic forces on panel and connections

4. Design forces for the brackets

5. Design forces for the rods

Calculations and Discussion   Code Reference

1. **Strength design load combinations**   §2.3.2

For design of the panel connections to the building, the applicable strength design load combinations are

a) $1.2D + 1.0E + \frac{1}{2}, L = 1.2D + 1.0E$ (Comb. 5)

where, with $E = \rho Q_E + 0.2\ S_{DS}D$ (Eq 12.4-1)

$$= 1.0Q_E + 0.2(1.0)D$$

$$= Q_E + 0.2D$$

As $S_{DS} = 1.0g$, the equation reduces to

$1.4D + Q_E$, for Q_E and D with same signs and type of load action.

b) $0.9D + 1.0E$ (Comb. 7)

with $E = \rho Q_E + 0.2\ S_{SD}D$ (Eq 12.4-1)

$$= Q_E + 0.2D$$

As $S_{DS} = 1.0g$, the equation reduces to

$1.1D + Q_E$, for Q_E and D with same signs.

This combination need not be considered since it is less than $1.4D + Q_E$.

c) $0.9D + 1.0Q_E$ (Comb. 7)

with $E = \rho Q_E - 0.2\ S_{SD}D$ (Eq 12.4-2)

$$= Q_E - 0.2D$$

As $S_{DS} = 1.0g$, the equation reduces to

$0.70D + Q_E$, for Q_E and D with opposite signs.

This combination need not be considered because the rod connections resist only the Q_E axial load, and the bracket connections have shear resistance capacity independent of the direction of the Q_E shear load: for example, upward resistance is equal to downward resistance. Therefore, this load combination is satisfied by $1.4D + Q_E$ for Q_E and D with the same signs.

In the seismic load combinations, Q_E is the load action on the connection due to the lateral load F_p applied either in-plane or out-of-plane at the panel center-of-mass per §13.3.

| **2.** | ## Lateral seismic force at center-of-mass C of panel |

Section 13.5.3, Item d., requires that the connection seismic load actions be determined by the force F_p given by §13.3.1 applied to the center-of-mass of the wall panel. The values of R_p and a_p are given in Table 13.5-1 for the body and fasteners of the connection elements.

To represent the average seismic acceleration on the panel, F_p will be determined as the average of the F_p values for the upper bracket elevation level, z_u, and for the lower rod elevation clevations, z_L. For the higher story levels of the building, this average F_p would be essentially equal to the F_p value using $z = z_c$ at the panel center-of-mass elevation. However, this use of elevations $z = z_c$ may not be valid for the lower story levels because of the limitation of

$$F_p \geq 0.3\, S_{DS}I_p W_p \tag{Eq 13.3-3}$$

With the given values of $S_{DS} = 1.0$, and $I_p = 1.0$

$$F_p = \frac{0.4 a_p S_{DS} I_p}{R_p}\left(1 + 2\frac{z}{h}\right) W_p \tag{Eq 13.3-1}$$

$$F_p \geq 0.3\, S_{DS}I_p W_p = 0.3(1.0)(1.0)W_p = 0.3 W_p \tag{Eq 13.3-3}$$

$$a_p = 1.0 \text{ and } R_p = 2.5, \text{ for body of connection} \tag{T 13.5-1}$$

$$W_p = \text{weight panel} = 14.4 \text{ kips}$$

At upper bracket connection level

$$z = z_u = 47 \text{ ft}$$

$$F_{pU} = \frac{0.4(1.0)}{2.5}\left[1 + 2\left(\frac{47}{60}\right)\right] W_p$$

$$= 0.411 W_p > 0.3\, S_{DS}I_p W_p = 0.3 W_p \dots o.k.$$

At lower rod connection level

$$z = z_L = 37 \text{ ft}$$

$$F_{pL} = \frac{0.4(1.0)}{2.5}\left[1 + 2\left(\frac{37}{60}\right)\right] W_p$$

$$= 0.357 W_p > 0.3 W_p \dots o.k.$$

The required average, $F_p = \dfrac{F_{pU} + F_{pL}}{2} = \dfrac{(0.411 + 0.357)}{2} W_p$

$$= 0.384 W_p = 0.384(14.)$$

$$= 5.53 \text{ kips}$$

This force is applied at the panel centroid C and acts horizontally in either the out-of-plane or the in-plane direction.

3. Combined dead and seismic forces on panel and connections   §13.5.2

There are two seismic load conditions to be considered: out-of-plane and in-plane. These are shown below as concentrated forces. In this example, Combination 5 of §2.3.2, $1.2D + Q_E$, is the controlling load combination.

a. Dead load, seismic out-of-plane, and vertical seismic forces

Panel connection reactions due to factored dead load, out-of-plane seismic forces, and vertical seismic forces are calculated as follows:

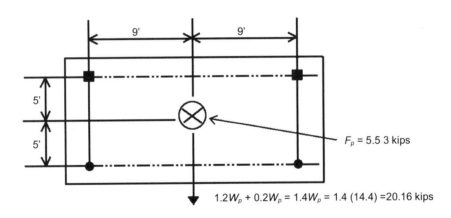

Each bracket and rod connection takes the following axial load due to the out-of-plane force F_p at center-of-mass

$$P_B + P_R = \frac{F_p}{4} = \frac{5.53}{4} = 1.38 \text{ kips}$$

where P_B is the bracket force and P_R is the rod force.

Each bracket takes the following downward in-plane shear force due to vertical loads

$$V_B = \frac{1.4W_p}{2} = \frac{20.16}{2} = 10.08 \text{ kips}$$

Note that each rod, because it carries only axial forces, has no in-plane, dead, or seismic loading.

b. Dead load, seismic in-plane, and vertical seismic forces

Panel connection reactions due to factored dead load, in-plane seismic forces, and vertical seismic forces are calculated as

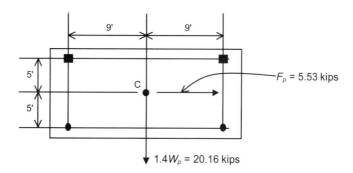

Each bracket takes the following in-plane horizontal shear force due to lateral seismic load

$$H_B = \frac{F_p}{2} = \frac{5.53}{2} = 2.77 \text{ kips}$$

Each bracket takes the following upward or downward shear force due to the reversible lateral seismic load

$$F_B = \frac{5(F_p)}{18} = \frac{5(5.53)}{18} = \pm 1.54 \text{ kips}$$

Each bracket takes the following downward force due to vertical loads:

$$R_B = \frac{1.4W_p}{2} = \frac{20.16}{2} = 10.08 \text{ kips}$$

Under the in-plane seismic loading, each rod carries no force.

4. Design forces for the brackets

a. Body of connection

Under §13.5.3 and Table 13.5.1 the body of the connection must be designed for $a_p = 1.0$ and $R_p = 2.5$. These are the a_p and R_p values used for the determination of F_p.

Therefore, there is no need to change the load actions due to this force.
The bracket must be designed to resist the following sets of load actions.

$$P_B = \pm 1.38 \text{ axial load together with}$$

$$V_B = 10.08 \text{ kips downward shear}$$

and

$$H_B = \pm 2.77 \text{ kips horizontal shear together with}$$

$$F_B + R_B = 1.54 + 10.08 = 11.62 \text{ kips downward shear}$$

b. Fasteners

Under §13.5.3, Item d., and Table 13.5.1, fasteners must be designed for $a_p = 1.25$ and $R_p = 1.0$. Thus, it is necessary to multiply the F_p load actions by $(1.25)(2.5) = 3.125$ because these values were based on $a_p = 1.0$ and $R_p = 2.5$. Fasteners must be designed to resist

$$(3.125)\, P_B = 3.125(1.38) = 4.31 \text{ kips axial load together with}$$

$$V_B = 10.08 \text{ kips downward shear}$$

and

$$3.125 H_B = 3.125(2.77) = 8.66 \text{ kips horizontal shear together with}$$

$$3.125 F_B + R_B = 3.125(1.54) + 10.08 = 14.89 \text{ kips downward shear}$$

5. Design forces for the rods

a. Body of connection

The body of the connection must be designed to resist a force based on $a_p = 1.0$ and $R_p = 2.5$

$$P_R = 1.39 \text{ kips axial load}$$

| **b.** | **Fasteners** |

Fasteners in the connecting system must be designed to resist a force based on $a_p = 1.25$ and $R_p = 1.0$

$$(3.125)P_R = 3.125(1.38) = 4.31 \text{ kips axial load}$$

Example 43
Beam Horizontal Tie Force   §12.1.3

This example illustrates use of the beam inter-connection requirement of §12.1.3. The requirement is to ensure that important parts of a structure are "tied together."

Find the minimum required tie capacity for the connection between the two simple beams shown in the example below.

The following information is given.

Seismic Design Category D

$S_{DS} = 1.0$

Dead Load D = 6 kip/ft

Live Load L = 4 kip/ft

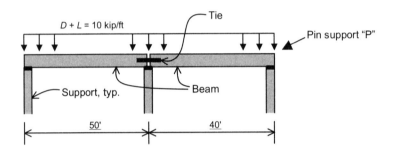

| **1.** | Determine tie force |
| **2.** | Determine horizontal support force at "P" |

Calculations and Discussion   **Code Reference**

1. **Determine tie force**

Requirements for ties and continuity are specified in §12.1.3. For this particular example, it is required to determine the "tie force" for design of the horizontal tie interconnecting the two simply supported beams. This force is designated as F_p, given by the greater value of

$$F_p = 0.133\, S_{DS} w_p$$

or

$$F_p = 0.05 w_p$$

where w_p is the weight of the smaller (shorter) beam

$$w_p = 40 \text{ ft } (D) = 40(6) = 240 \text{ kips}$$

Example 43 ■ *Beam Horizontal Tie Force* *§12.1.3*

For $S_{DS} = 1.0$, the controlling tie force is

$$F_p = 0.133(1.0)(240) = 31.9 \text{ kips}$$

---

2. Determine horizontal support force at "P"

Section 12.1.4 requires a horizontal support force for each beam equal to 5 percent of the dead plus live load reaction. Given a sliding bearing at the left support of the 40-foot beam, the required design force at the pin support "P" is

$$H = 0.05(6 \text{ klf} + 4 \text{ klf})\left(\frac{40}{2}\right) = 10 \text{ kips}$$

Example 44
Collector Elements   §12.10.2

Collectors "collect" forces and carry them to vertical shear-resisting elements. Collectors are sometimes called drag struts. The purpose of this example is to show the determination of the maximum seismic force for design of collector elements. In the example below, a tilt-up building, with special reinforced concrete shear walls and a panelized wood roof, has a partial interior shear wall on Line 2. A collector is necessary to "collect" the diaphragm loads tributary to Line 2 and bring them to the shear wall.

The following information is given.

Occupancy Category I

Seismic Design Category D

R = 5.5
Ω_o = 2.5
I = 1.0
S_{DS} = 1.20
Roof dead load = 15 psf
Wall height = 30 ft, no parapet
Wall weight = 113 psf

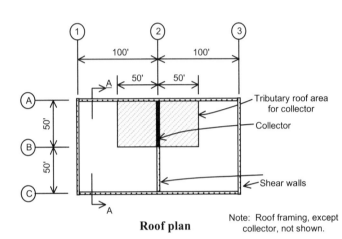

Roof plan

Note: Roof framing, except collector, not shown.

By inspection, for the one-story shear wall building, Equation 12.8-2 will govern.

$$\text{Base shear} = V = \frac{S_{DS}}{R} W = 0.218W \qquad \text{(Eq 12.8-2)}$$

W = structure weight above one half h_1

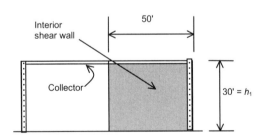

Elevation Section A-A

Determine the following.

1. Collector unfactored force at tie to wall

2. Special seismic load of §12.4.3.2 at tie to wall

Example 44 ■ *Collector Elements*   §12.10.2

Calculations and Discussion   **Code Reference**

1. Collector unfactored force at tie to wall   §12.10.2

The seismic force in the collector is made up of two parts: 1) the tributary out-of-plane wall forces, and 2) the tributary roof diaphragm force. The panelized wood roof has been determined to be flexible; thus the tributary roof area is taken as the 100-foot by 50-foot area shown on the roof plan above. Seismic forces for collector design are determined from Equation 12.10-1 used for diaphragm design. This equation reduces to the following for a single story structure.

$$F_{p1} = \frac{F_1}{W_1} w_{p1}$$

$$F_{p1} \, \text{max} = 0.4 \, S_{DS} I W_{p1} = 0.40 W_{p1}$$

$$F_{p1} \, \text{min} = 0.2 \, S_{DS} I W_{p1} = 0.2 W_{p1}$$

F_1 = design force at roof
W_1 = structure weight above one half $h_1 = W$
w_{p1} = weight tributary to the collector element

giving:

$$F_{p1} = \frac{V}{W} w_{p1} = 0.218 W_{p1}$$

w_{p1} = tributary roof and out-of-plane wall weight

$$w_{p1} = 15 \, \text{psf}(100)(50) + 113 \, \text{psf}\left(\frac{30}{2}\right)(100) = 75,000 + 169,500 = 244.5 \, \text{kips}$$

$$\therefore F_{p1} = 0.218(244.5) = 53.3 \, \text{kips}.$$

Note: This force corresponds to the diaphragm design forces calculated using §12.10.1. These forces are compared to the diaphragm shear strength; including the shear strength of connection between the diaphragm and collector. The design of the collector and its connections requires that the axial forces be amplified as shown below.

2. Special seismic load of §12.4.3.2 at tie to wall   §12.10.2

Given the force F_{p1} specified by Equation 12.10-1, the collector elements, splices, and their connections to resisting elements shall have the design strength to resist the earthquake loads as defined in the Special Load Combinations of §12.4.3.2.

The governing load combination is

$$1.2\,D + 0.5L + E_m \qquad\qquad §2.3.2 \text{ (Comb. 5)}$$

where

$$E_m = \Omega_o Q_E + 0.2\,S_{DS}D \qquad\qquad \text{(Eq 12.4-5)}$$

Here, Q_E is the horizontal collector design force $F_{p1} = 53.3$ kips, and

$$\Omega_o Q_E = 2.5(53.3) = 133.25 \text{ kips axial tension and compression load}$$

$$0.2\,S_{DS}D = 0.2(1.2)D = 0.24D \text{ vertical load}$$

The strength design of the collector and its connections must resist the following load components.

$$\Omega_o Q_E = 2.5(53.3) = 133.25 \text{ kips axial tension and compression load}$$
and vertical downward load equal to

$$1.2D + 0.5L + 0.24\,D = 1.44\,D + 0.5\,L$$

Assume tributary width for D and L is 16'.

with $\quad D = (50 \text{ ft} + 50 \text{ ft})(16 \text{ ft})(15 \text{ psf}) = 24{,}000 \text{ lb}$

$\quad L = (50 \text{ ft} + 50 \text{ ft})(16 \text{ ft})(20 \text{ psf}) = 32{,}000 \text{ lb}$

The resulting total factored vertical load is

$$1.44(24{,}000) + 0.5(32{,}000) = 50{,}560 \text{ lb}$$

which is applied as a uniform distributed load w = 50,560 lb/50 ft = 1011 plf on the 50-foot length of the collector element.

Commentary

Note that §12.4.3.1 specifies that the term $\Omega_o Q_E$ in Equation 12.4-7 need not exceed the maximum force that can be delivered by the lateral-force-resisting system as determined by rational analysis. For example, the overturning moment capacity of the shear wall can limit the required strength of the collector and its connection to the shear wall.

Example 45
Out-of-Plane Wall Anchorage of Concrete or
Masonry Walls to Flexible Diaphragms　§12.11.2 and 12.11.2.1

For the tilt-up wall panel shown below, the seismic force required for the design of the wall anchorage to the flexible roof diaphragm is to be determined. This will be done for a representative 1-foot width of wall.

The following information is given.

Occupancy Importance Category I

Seismic Design Category D

I　　= 1.0
S_{DS}　= 1.0
Panel thickness = 8 in
Normal weight concrete (150 pcf)

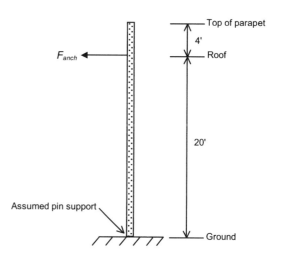

Determine the following.

| 1. | Design criteria |
| 2. | Wall anchorage force |

Calculations and Discussion　　　　　　　　　　　　　　　Code Reference

1. Design criteria　　　　　　　　　　　　　　　　　　　§12.11.2.1

Because of the frequent failure of wall/roof ties in past earthquakes, the code requires that the force used to design wall anchorage to flexible diaphragms be greater than that given in §12.11.2.1 for the design of the wall panel sections. The following equation is to be used to determine anchor design forces, with minimum limit given in §12.11.2.

$$F_p = 0.8 \, S_{DS} I_E w_w \qquad\qquad\qquad \text{(Eq 12.11-1)}$$

$$\geq 400 \, S_{DS} I \text{ lb/ft}$$

$$\geq 280 \text{ lb/ft of wall}$$

where w_w is the weight of a 1-foot width of wall that is tributary to the anchor.

| **2.** | **Wall anchorage force** |

The tributary wall weight is one-half of the weight between the roof and base *plus* all the weight above the roof.

$$w_w = 150\left(\frac{8}{12}\right)(4\text{ ft} + 10\text{ ft})(1\text{ ft}) = 1400\text{ lb/ft}$$

For the given values of $S_{DS} = 1.0$ and $I = 1.0$, Equation 12.11-1 gives

$$F_p = 0.8(1.0)(1.0)w_p = 0.8w_p = 0.8(1400)$$

$$= 1120\text{ lb/ft} > 400(1.0)(1.0) = 400\text{ lb/ft} \dots o.k.$$

$$> 280\text{ lb/ft} \dots o.k.$$

$$\therefore F_{anch} = F_p = 1120\text{ lb/ft}$$

This is the Q_E load in the seismic load combinations.

Example 46 ■ *Wall Anchorage to Flexible Diaphragms* **§12.11.2.1**

Example 46
Wall Anchorage to Flexible Diaphragms §12.11.2.1

This example illustrates use of the allowable stress design procedure for the design of steel and wood elements of the wall anchorage system in a building with a flexible roof diaphragm.

The drawing below shows a tilt-up wall panel that is connected near its top to a flexible roof diaphragm. The anchorage force has been calculated per §12.11.2.1 as F_{anch} = 1680 lb/ft. The wall anchorage connections to the roof are to be provided at 4 feet on center.

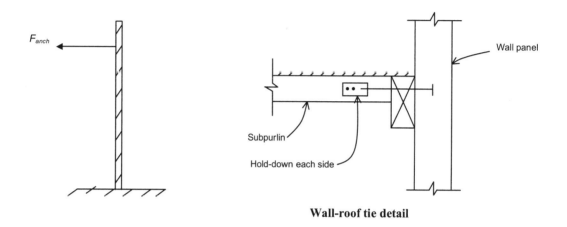

Wall-roof tie detail

Determine the strength design requirements for the following.

1. Design force for premanufactured steel anchorage element

2. Design force for wood subpurlin tie element

Calculations and Discussion **Code Reference**

1. Design force for premanufactured steel anchorage element.

The task is to design the steel anchorage elements (i.e., hold-downs) that connect the tilt-up wall panel to the wood subpurlins of the roof diaphragm. The anchorage consists of two hold-down elements, one on each side of the subpurlin.
The manufacturer's catalog provides allowable capacity values for earthquake loading for a given type and size of hold-down element.

The steel hold-down elements of the anchorage system resist only the axial anchorage load and there are no dead or live load effects.

For the 4-foot spacing, the strength design axial load is

$$E = Q_E = P_E = F_{anch}(4) = (1680)(4) = \pm 6720 \text{ lb}$$

This example, uses the ASD load combinations of §2.4, where the applicable seismic load combinations permit $0.7E$ to be resisted with an increase in allowable stress based on duration (i.e., the C_d duration factor for wood).

The allowable stress design axial load requirement for each pair of hold-down elements is

$$0.7E = 0.7P_E \, 0.7(6720) = \pm 4800 \text{ lb}$$

From the manufacturer's catalog, select a hold-down element having a capacity of at least

$$\frac{4800}{2} \text{ lb} = 2400 \text{ lb}$$

The hold-down detail must provide both tensile and compressive resistance for this load.

Whenever hold-downs are used in pairs, as shown in the wall-roof tie detail above, the through-bolts in the subpurlin must be checked for double shear bearing. Also, the paired anchorage embedment in the wall is likely to involve an overlapping pull-out cone condition in the concrete: refer to ACI 318 Appendix D for design requirements. When single-sided hold-downs are used, these must consider the effects of eccentricity. Generally, double hold-downs are preferred, but single-sided hold-downs are often used with all eccentricities fully considered.

2. | Design force for wood subpurlin tie element

The strength design axial load on the wood element of the wall anchorage system is

$$P_E = (1680)(4) = \pm 6720 \text{ lb}$$

Using the seismic load combinations of §2.4, select the wood element such that the allowable capacity of the element, for the combined bending and axial stress including dead and live load effects, can support a $\pm$ axial load of

$$0.7P_E = 0.7(6720) = 4800 \text{ lb applied at the anchored end.}$$

Example 46 ■ *Wall Anchorage to Flexible Diaphragms* §12.11.2.1

Commentary

For comparison, the forces acting on wood, concrete, and steel elements are shown below. For wood, the load is divided by the duration factor C_d of 1.0 to permit comparison. For steel, the load is increased by 1.4 per §12.11.2.2.

Material	F_p/C_d	ASD
Wood	$\dfrac{0.8S_{DS}IW}{1.6} = 0.5\ S_{DS}IW$	$(0.35\ S_{DS}IW)$
Concrete	$0.8\ S_{DS}IW$	N/A
Steel	$1.4(0.8\ S_{DS}IW) = 1.12\ S_{DS}IW$	$(0.78\ S_{DS}IW)$

Example 47

Determination of Diaphragm Force F_{px} : Lowrise   §12.10.1.1

This example illustrates determination of the diaphragm design force F_{px} of
Equation 12.10-1, for the design of the roof diaphragm of a single-story building.

A single-story tilt-up building with special reinforced concrete shear walls and a panelized
wood roof is shown below. This type of roof construction can generally be shown to
behave per flexible diaphragm assumptions.

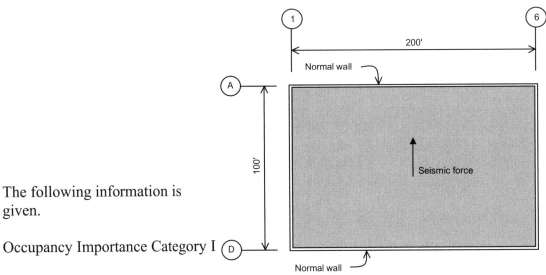

Roof plan

The following information is
given.

Occupancy Importance Category I

Seismic Design Category D

I = 1.0
S_{DS} = 1.0
R = 5.0
ρ = 1.0
Diaphragm weight = 15 psf
Wall weight = 80 psf

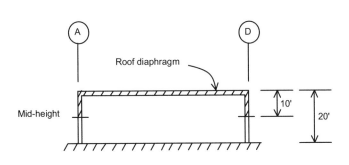

Elevation through building

Find the following.

| **1.** | **Diaphragm design force at the roof** |

Example 47 ■ *Determination of Diaphragm Force F_{px}: Lowrise*   §12.10.1.1

Calculations and Discussion   **Code Reference**

1. **Diaphragm design force at the roof**   §12.10.1.1

§12.10.1.1 requires that the design seismic force for diaphragms be determined by

$$F_{px} = \frac{\displaystyle\sum_{i=x}^{n} F_i}{\displaystyle\sum_{i=x}^{n} w_i} = w_{px}$$ (Eq 12.10-1)

with limits of

$$0.2\, S_{DS} I w_{px} \le F_{px} \le 0.4\, S_{DS} I w_{px}$$

which for

$$S_{DS} = 1.0 \text{ and } I = 1.0$$

are

$$0.2 w_{px} \le F_{px} \le 0.4 w_{px}$$

For a short period single story building, Equation 12.10-1 becomes (see commentary below for derivation)

$$F_{px} = F_{p1} = \frac{S_{DS} I}{R} w_{p1}$$

with the given values of $S_{DS} = 1.0$, $R = 5.0$

and, for a 1-foot-wide strip

$$w_{p1} = \text{ weight of diaphragm} + \text{weight of one-half height of normal walls}$$

$$= 100(15) + 2(10)(80) = 3100 \text{ lb/ft}$$

$$F_{p1} = \frac{(1.0)(1.0)}{5.0} w_{px} = 0.2\, w_{p1} = 0.2(3100) = 620 \text{ lb/ft}$$

Check limits: $0.2 w_{px} < 0.2 w_{p1} < 0.4 w_{px}$ . . . *o.k.*

Note that the redundancy factor of ρ is to be applied to the Q_E load actions due to F_{p1} (such as chord forces and diaphragm shear loads in the diaphragm).

Commentary

1. The weight w_{px} includes the weight of the diaphragm plus the tributary weight of elements normal to the diaphragm that are one-half story height below and above the diaphragm level. Walls parallel to the direction of the seismic forces are usually not considered in the determination of the tributary roof weight because these walls do not obtain support, in the direction of the force, from the roof diaphragm.

2. The single-story building version of Equation 16-65 is derived as follows.

$$F_{px} = \frac{\sum\limits_{i=x}^{m} F_i}{\sum\limits_{i=x}^{n} w_i} w_{px} \qquad \text{(Eq 12.10-1)}$$

$$F_x = C_{vx}V = \frac{V\, w_x h_x^k}{\sum\limits_{i=1}^{n} w_i h_i^k} \qquad \text{(Eq 12.8-11)}$$

where $C_{vx} = \dfrac{w_x h_x^k}{\sum\limits_{i=1}^{n} w_i h_i^k}$ for short period of < 0.5 sec ($k = 1.0$). $\qquad$ (Eq 12.8-12)

For a single-story building,

$$i = 1, \ x = 1, \ \text{and } n = 1$$

$$\sum_{i=1}^{1} w_i = W$$

and Equation 12.8-11 gives

$$F_1 = \frac{w_1 h_1}{w_1 h_1} \, V = V$$

where

$$V = C_s W = \frac{S_{DS} I}{R} W$$ (Eq 12.8-1 and 12.8-2)

Finally, for the single story building, Equation 12.10-1 is

$$F_{p1} = \frac{F_1}{W} w_{p1} = \frac{V}{W} w_{p1} = \frac{S_{DS} I}{R} w_{p1}$$

Example 48
Determination of Diaphragm Force F_{px}: Highrise   §12.10.1

This example illustrates determination of the diaphragm design force F_{px} of Equation 12.10-1 for a representative floor of a multi-story building.

The nine-story moment frame building shown below has the tabulated design seismic forces F_x. These were determined from Equations 12.8-11 and 12.8-12, the design base shear.

The following information is given.

Seismic Design Category D

$W = 3,762$ kips
$C_s = 0.06215$
$S_{DS} = 1.0$
$\rho = 1.3$
$I = 1.0$
$T = 1.06$ sec
$V = C_s W = 233.8$ kips
$k = 2$ for Eq 12.8-12

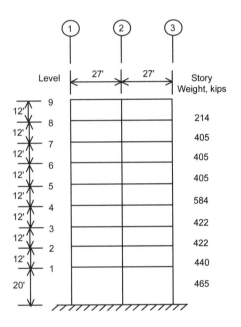

Level x	h_x (ft)	h_x^k	w_x kips	$w_x h_x^k$	$C_{vx} = \dfrac{w_x h_x^k}{\sum w_i h_i^k}$	$F_x = C_{vx}V$	$\dfrac{F_x}{w}$
9	116	13456	214	2879584	0.153	35.8	0.167
8	104	10816	405	4380480	0.233	54.4	0.134
7	92	8464	405	3427920	0.183	42.8	0.106
6	80	6400	405	2592000	0.138	32.3	0.079
5	68	4624	584	2700416	0.144	33.7	0.058
4	56	3136	422	1323392	0.071	16.6	0.039
3	44	1936	422	816992	0.044	10.3	0.024
2	32	1024	440	450560	0.024	5.6	0.013
1	20	400	465	186000	0.010	2.3	0.005
Totals:			3,762	18,757,344		233.8	

Calculations and Discussion	**Code Reference**

1. **Diaphragm force at Level 7**   §12.10-1

Seismic forces on the floor and roof diaphragm are specified in §12.10-1. The following equation is used to determine the diaphragm force F_{px} at Level x

$$F_{px} = \frac{\sum_{i=x}^{n} F_i}{\sum_{i=x}^{n} w_i} w_{px}$$ (Eq 12.10-1)

Section 12.10.1.1 also has the following limits on F_{px}

$$0.2\, S_{DS} I w_{px} \leq F_{px} \leq 0.4 S_{DS} I w_{px}$$

For Level 7, $x = 7$

$$F_{p7} = \frac{(42.8 + 54.4 + 35.8)(405)}{(405 + 405 + 214)} = (0.130)(405) = 52.6 \text{ kips}$$

Check limits:

$0.2\, S_{DS} I w_{px} \quad = 0.2 w_{px}$

$\qquad\qquad\qquad = 0.2(405) = 81.1 \text{ kips} > 52.6 \text{ kips} \ldots$ **not** *o.k.*

$0.4\, S_{DS} I w_{px} \quad = 0.4 w_{px}$

$\qquad\qquad\qquad = 0.4(405) = 162.0 \text{ kips} > 52.6 \text{ kips} \ldots o.k.$

∴ $F_{p7} = 81.1$ kips…minimum value $(0.2\, S_{DS} I w_{px})$ governs.

Note that the redundancy factor, in this example $\rho = 1.3$, is to be applied to the load Q_E due to F_{px} (such as chord forces and floor-to-frame shear connections). Also note that Equation 12.10-1 will always govern for the design of the diaphragm versus Equation 12.8-12.

Example 49
Building Separations   §12.12.3

Building separations are necessary to prevent or reduce the possibility of two adjacent structures impacting during an earthquake. Requirements for building separations are given in §12.12.3. In this example, the static displacements δ_{xe} due to the prescribed lateral forces of §12.8 and information about each structure are given below. Note that the displacements given are at the plan view edges of the building.

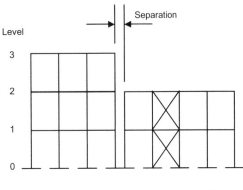

	Structure *1*		Structure *2*	
Level *x*	δ_{xe}		Level	δ_{xe}
3	1.38 in.		—	—
2	1.00		2	0.75 in
1	0.47		1	0.35
0	0		0	0
$R = 8$			$R = 6$	
$C_d = 5.5$			$C_d = 5$	

Structure 1   Structure 2

Find the required separations for the following situations.

1. Separations within the same building

2. Separation from an adjacent building on the *same* property

3. Separation from an adjacent building on *another* property

Calculations and Discussion   Code Reference

1. **Separation within the same building**   §12.12.3

Expansion joints are often used to break a large building or an irregular building into two or more parts above the foundation level. This effectively creates separate structures within the same building. The code requires that the structures be separated by the amount δ_{MT}

where

$$\delta_{MT} = \delta_{M1} + \delta_{M2}$$

$$\delta_{M1} = \text{maximum inelastic displacement of Structure 1}$$

$$\delta_{M2} = \text{maximum inelastic displacement of Structure 2}$$

Example 49 ▪ *Building Separations* §12.12.3

where

$$\delta_M = (\delta_x)_{max} = \frac{C_d \delta_{xe}}{I} \qquad \text{(Eq 12.8-1)}$$

The required separation is determined in the following two steps.

a. **Determine inelastic displacements of each structure** §12.8.6

To determine the minimum separation bctwccn parts of the same building that are separated by an expansion joint, the <u>maximum</u> inelastic floor displacements δ_x must be determined for each structure. These are at level, $x = 2$

For Structure 1

$$\delta_{M1} = \frac{C_d \delta_{21}}{I} = \frac{5.5(1.0)}{1.0} = 5.5 \text{ in} \qquad \text{(Eq 12.8-15)}$$

For Structure 2

$$\delta_{M2} = \frac{C_d \delta_{22}}{I} = \frac{5.0(0.75)}{1.0} = 3.75 \text{ in} \qquad \text{(Eq 12.8-15)}$$

b. **Determine the required separation** §12.12.3

The required separation is determined from the individual maximum inelastic displacements of each structure as

$$\delta_{MT} = \delta_{M1} + \delta_{M2} = 5.5 + 3.75 = 9.25 \text{ in}$$

2. **Separation from an adjacent building on the *same* property**

If Structures 1 and 2 above are adjacent, individual buildings on the same property, the solution is the same as that shown above in Step 1. The code makes no distinction between an "internal" separation in the same building and the separation required between two adjacent buildings on the same property.

$$\delta_{MT} = 9.25 \text{ in}$$

| 3. | **Separation from an adjacent building on *another* property**   §12.12.3 |

If Structure 1 is a building under design and Structure 2 is an existing building on an adjoining property, we would generally not have information about the seismic displacements of Structure 2. Often even basic information about the structural system of Structure 2 may not be known. In this case, separation must be based only on information about Structure 1. The largest elastic displacement of Structure 1 is $\delta_{3e} = 1.38$ inches and occurs at the roof (Level 3). The maximum inelastic displacement is calculated as

$$\delta_M = \frac{C_d \delta_{3e}}{I} = \frac{5.5(1.38)}{1.0} = 7.59 \text{ in} \qquad \text{(Eq 12.8-15)}$$

Structure 1 must be set back 7.59 inches from the property line, unless a smaller separation is justified by a rational analysis based on maximum ground motions. Such an analysis is difficult to perform, and is generally not required except in very special cases.

| 4. | **Seismic separation between adjacent buildings** |

SEAOC recommends the following seismic separation between adjacent buildings.

$$\delta = \sqrt{(\delta_{M1})^2 + (\delta^2{}_{M2})}$$

Example 50 ■ *Flexible Nonbuilding Structure* §15.5

Example 50
Flexible Nonbuilding Structure §15.5

A tall steel bin tower is supported by a concrete foundation. The tower sits on symmetrically braced legs

The following information is given.

Seismic Design Category D
      Weight of tower and maximum
      normal operating contents = 150 kips
Occupancy Category III
Site Class D
    I = 1.25 per Table 11.5-1
    S_s = 1.70, S_1 = 0.65
    S_{DS} = 1.20, S_{DI} = 0.65

The stiffness of the supporting
tower is 8.30 kip/in

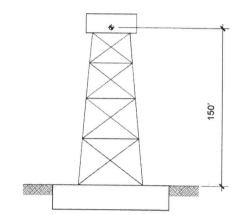

Determine the following.

1. **Period of vibration**

2. **Design base shear**

3. **Vertical distribution of seismic forces**

4. **Overturning moment at base**

Calculations and Discussion
Code Reference

1. **Period of vibration.**

For calculation purposes, the mass is assumed to be located at the top of the tower. The period must be determined by §15.4.4.

$$T = 2\pi\sqrt{\frac{m}{k}} = 2\pi\sqrt{\frac{150 \text{ kips}/(386 \text{ kip/in/sec}^2)}{8.30 \text{ kip/in}}} = 1.36 \text{ sec}$$

Because the period is greater than .06 second, the vessel does not qualify as a rigid nonbuilding structure and thus is considered flexible. See §15.4.2.

It should be noted that the value of the period, T, should not be calculated using any of the approximate methods in §12.8.2.1, nor is it intended to be subject to the limitations presented in §12.8.2. This is because the approximate method presented is intended for buildings and is not applicable to structural systems that differ significantly from typical building configurations and characteristics. Refer to Section C109.1.4 of the 1999 SEAOC Blue Book for further discussion.

2. Design base shear

The design base shear for nonbuilding structures is calculated from the same expressions as for buildings. These are given in §12.8.1. In addition, nonbuilding structures, such as the vessel, must also satisfy the requirements of §15.5.

$$V = C_s W = 0.217\,(150 \text{ kips}) = 32.55 \text{ kips}$$

where

$$C_s = \frac{S_{DS}}{R/I} = 0.50 \qquad\qquad\qquad \text{(Eq 12.8-2)}$$

$S_{DS} = 1.2$
$R\ \ = 3.0$ T 15.4-2
$\Omega_o = 2.0$ T 15.4-2
$C_D = 2.5$ T 15.4-2
$I\ \ = 1.25$ T 11.5-1

The value of C_s computed in accordance with Eq. 12.8-2 need not exceed

$$C_s = \frac{S_{D1}}{(R/I)T} = 0.199 \qquad\qquad\qquad \text{(Eq 12.8-2)}$$

where

$S_{D1} = 0.65$
$R\ \ = 3.0$
$I\ \ = 1.25$
$T\ \ = 1.36 \text{ sec}$

But C_s shall not be taken less than

$$C_s = 0.01 \qquad\qquad\qquad\qquad\qquad \text{(Eq 12.8-5)}$$

where

$S_{DS} = 1.20$
$I\ \ = 1.25$

Example 50 ■ *Flexible Nonbuilding Structure*   §15.5

Note that for this tower, because the 1-second spectral response S_1 is equal to 0.65, ($S_1 \geq 0.60g$), the value of the seismic response coefficient C_s shall not be taken as less than

$$C_s = \frac{0.5 S_1}{(R/I)} = 0.135$$ (Eq 12.8-6)

where

$S_1 = 0.65$
$R = 3.0$
$I = 1.25$

Thus: $C_s = 0.199$ governs

Also note that if this tower (Occupancy Category II) were located on a site with mapped maximum considered earthquake spectral response acceleration at 1-second period S_1, equal to or greater than 0.75g, it would be assigned to SDC E (§11.6). Thus, the height would be limited to 100 ft per Table 15.4-2.

Example 51
Lateral Force on Nonbuilding Structure   §15.0

A nonbuilding structure with a special reinforced concrete moment frame (SRCMF) supports some rigid aggregate storage bins. Weights W_1 and W_2 include the maximum normal operating weights of the storage bins and contents as well as the tributary frame weight. See §15.4.1.1 and Table 11.5.1

The following information is given.

Occupancy Importance Category I
$I = 1.0$

Site Class D
$S_{MS} = 2.0, S_s = 2.0$
$S_{M1} = 1.5, S_1 = 1.0$
$S_{DS} = 1.33$
$S_{D1} = 1.00$
$T = 2.0$ sec
$W = 300$ kips

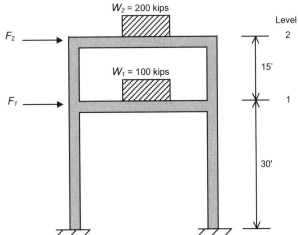

Determine the following.

1. **Design base shear**

2. **Vertical distribution of seismic forces**

Calculations and Discussion   **Code Reference**

1. **Design base shear.**   **§15.4**

Because this is a flexible structure, (i.e., the period T > 0.06 sec, see §15.4.2, and the structure is similar to a building, see §15.4.1) the general expressions for design base shear given in §12.4 and §15.4 must be used. Note that an intermediate reinforced concrete moment frame (IRCMF) building structure is <u>not</u> permitted for SDC D, E, or F per Table 15.4.1. Also note that the value for R is 8 for normal design of an SRCMF.

The total base shear in a given direction is determined from

$$V = C_s W \qquad\qquad\qquad\text{(Eq 12.8-1)}$$

where

$$C_s = \frac{S_{DS}}{(R/I)} = \frac{(1.33)}{(8.0)/(1.0)} = 0.166 \qquad\qquad\text{(Eq 12.8-2)}$$

Example 51 ■ *Lateral Force on Nonbuilding Structure* §15.0

where

$$S_{DS} = 1.33$$
$$R = 8$$
$$I = 1.0$$

The value of C_s computed in accordance with Equation 12.8-2 need not exceed

$$C_s = \frac{S_{DI}}{(R/I)T} = \frac{(1.0)}{(8/1.0)2.0} \text{ for } T \leq T_L = 0.063 \qquad \text{(Eq 12.8-3)}$$

where

$$S_{DI} = 1.0$$
$$R = 8.0$$
$$I - 1.0$$
$$T = 2.0$$

Check $T \leq T_L = > T_L = 12.0$ sec (Region 1, F 22-16)

The value of C_s shall not be taken less than

$$C_s = \frac{0.5S_1}{(R/I)} = \frac{(0.5)(1.0)}{\left(\dfrac{8}{1.0}\right)} = 0.063 \qquad \text{(Eq 12.8-6)}$$

where

$$S_1 = 1.0 \qquad \text{Note } S_1 \geq 0.6g$$
$$R = 8$$
$$I = 1.0$$
$$T = 2.0$$

Thus: $C_s = 0.063$ Equations 12.8-3 and 12.8-6 govern.

$$V = C_s W = (0.063)(300) = 18.9 \text{ kips} \qquad \text{(Eq 12.8-1)}$$

2. Vertical distribution of seismic forces §12.8-2

The design base shear must be distributed over the height of the structure in the same manner as that for a building structure.

$$F_x = C_{vx} V = C_{vx} (18.9 \text{ kips}) \qquad \text{(Eq 12.8-11)}$$

where

$$C_{vx} = \frac{w_x h_x^k}{\sum_{i=1}^{n} w_i h_i^k}$$

(Eq 12.8-11)

where

$k = 1.0$ for $T \le 0.50$ sec

and $k = 2.0$ for $T \ge 2.50$ sec

and $k =$ interpolate between 1 and 2.5 sec

Thus:

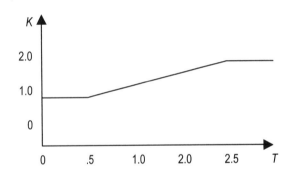

Now for $T = 2.0$ sec

$$k \quad = 1.0 + 1.0 \left(\frac{2.0 - 0.5}{2.5 - 0.5}\right) = 1.75$$

Example 51 ■ *Lateral Force on Nonbuilding Structure* §15.0

Story Shears ($k = 1.75$)

Level	Height h_x	h_x^k	Weight W_x	$W_x h_x^k$	C_{vx}	Story Force F_x	Story Shear V_x	$\bar{S}_a$
2	45	781.85	200	156369.45	0.803	15.17	15.17	0.076
1	30	384.56	100	38455.83	0.197	3.73	18.9	0.037
			300	194825.28	1.00		18.9	

Note: $k = 1.75$ $\qquad$ $C_{vx} = W_x h_x^k / 194825.28$

$\quad$ h_x in feet $\qquad\quad$ $F_x = C_{vx}(18.9)$

$\quad$ W_x in kips $\qquad\quad$ $\bar{S}_a = F_r / W_r \cong \Gamma\phi S_a$

$\qquad\qquad\qquad\qquad$ = effective story acceleration

Example 52
Rigid Nonbuilding Structure   §15.4.2

The code has special requirements for the determination of seismic forces for design
of rigid nonbuilding structures. In this example, rigid ore crushing equipment is supported
by a massive concrete pedestal and seismic design forces are to be determined.

The following information is given.

$S_{DS} = 1.33$
$I \quad = 1.0$
$T \quad = 0.02$ sec
$W_{EQUIPMENT} = 100$ kips
$W_{SUPPORT} = 200$ kips

Determine the following.

1. **Design base shear**

2. **Vertical distribution of seismic forces**

Calculations and Discussion   **Code Reference**

1. **Design base shear**   §15.4.2

For rigid nonbuilding structures, Equation 15.4-5 is used to determine design
base shear.

$$V = 0.3S_{DS}\,I\,W = 0.3\,(1.33)(1.0)\,W = 0.399W \qquad \text{(Eq 15.4-5)}$$

$$= 0.399\,(100 + 200) = 119.7 \text{ kips}$$

2. **Vertical distribution of seismic forces**

The force shall be distributed with height in accordance with §12.8.3

$$Fx = C_{vx}V = C_{vx}\,(119.7 \text{ kips}) \qquad \text{(Eq 12.8-11)}$$

Example 52 ■ *Rigid Nonbuilding Structure* §15.4.2

where

$$C_{vx} = \frac{w_x h_x^k}{\displaystyle\sum_{i=1}^{n} w_i h_i^k}$$

(Eq 12.8-12)

where

k = 1.0 for t ≤ 0.50 sec
T = 0.02 sec

Thus: k = 1.0

Story shears (*k* = 1.0)

Level	Height h_x	h_x^k	Weight W_x	$W_x h_x^k$	C_{vx}	Story Force F_x	Story Shear V_x	$\overline{S}_a$
2	30	30	100	3000	0.429	51.25	51.35	0.516
1	20	20	200	4000	0.571	68.45	119.7	0.342
			300	7000	1.00		119.7	

Note: h_x in feet $F_x = C_{vx}$ (119.7 kips)
        W_x in kips $\overline{S}_a = F_x / W_x \cong \Gamma \phi S_a$
        $C_{vx} = W_x h_x^k / \sum W_x h_x^k$ = effective story acceleration

Example 53
Tank With Supported Bottom   §15.7.6

A small liquid storage tank is supported on a concrete slab. The tank does not contain toxic or explosive substances.

The following information is given.

$S_{DS} = 1.20$
$I = 1.0$
W = Weight of tank and maximum normal operating contents
$\quad$ = 120 kips
t = 0.50 inch

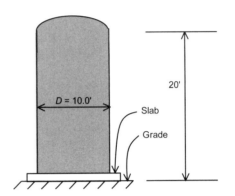

---

1. **Find the design base shear**   §15.7.6

Calculations and Discussion   **Code Reference**

---

1. The tank is a nonbuilding structure, and seismic requirements for tanks with supported bottoms are given in §15.7.6. This section requires that seismic forces be determined using the procedures of §15.4.2.

The period may be computed by other rational methods, similar to Example 51

$$T = 7.65 \times 10^{-6} \left(\frac{L}{D}\right)^2 \left(\frac{w \times D}{t}\right)^{1/2}$$

where

L = 20 ft
D = 10 ft
L/D = 20/10 = 2.0
w = W/L = 120,000 lb/20 = 6000 plf
t = 0.50 in
$\dfrac{wd}{t}$ = $\dfrac{6000(10)}{(0.50/12)}$ = 1,440,000

Example 53 ■ *Tank With Supported Bottom* §15.7.6

Now:

$$T = 7.65 \times 10^{-6} (2.0)^2 (1,440,000)$$
$$= 0.0367 \text{ sec} < 0.06 \ldots \text{rigid}$$

Thus, rigid nonbuilding structure, §15.4.2

The lateral force shall be obtained as follows

$$V = 0.3 \, S_{DS} \, IW = 0.36W \qquad \text{(Eq 15.4-5)}$$
$$= 0.36 \, (120) = 43.2 \text{ kips}$$

where

$$S_{DS} = 1.20$$
$$I = 1.00$$
$$W = 120 \text{ kips}$$

The design lateral seismic force is to be applied at the center-of-mass of the tank and its contents. Note that the center-of-mass of the contents and of the tank do not normally coincide. The distribution of forces vertically shall be in accordance with §12.8.3.

Commentary

The procedures above are intended for tanks that have relatively small diameters (less than 20 feet) and where the forces generated by fluid-sloshing modes are small. For large diameter tanks, the effects of sloshing must be considered. Refer to American Water Works Association Standard ANSI/AWWA D100 "Welded Steel Tanks for Water Storage," or American Petroleum Institute Standard 650, "Welded Steel Tanks for Oil Storage" for more detailed guidance.

Example 54
Pile Interconnections

IBC §1808.2.23.1

A two-story masonry bearing wall structure has a pile foundation. Piles are located around the perimeter of the building. The foundation plan of the building is shown below.

The following information is given.

Seismic Design Category D

$I\ = 1.0$

$S_{DS} = 1.0$

Pile cap size: 3 feet square by 2 feet deep

Grade beam: 1 foot 6 inches by 2 feet

Allowable lateral bearing = 200 psf

per foot of depth below natural grade,

for the very dense granular soil at the site.

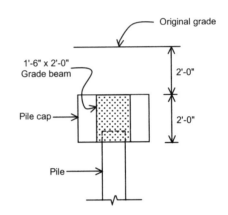

Section A-A: Typical pile cap

Pile Cap	Dead Load	Reduced Live Load	Seismic Q_E	
			N/S	E/W
3	46 kips	16 kips	14 kips	0
10	58	16	14	0

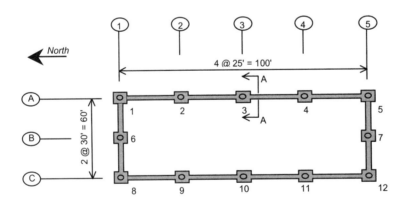

Foundation plan

Determine the following.

1. **Interconnection requirements**

2. **Interconnection force between pile caps 3 and 10**

3. **Required "tie" restraint between pile caps 3 and 10**

Example 54 ■ *Pile Interconnections*   *IBC §1808.2.23.1*

Calculations and Discussion	*Code Reference*

1. Interconnection requirements

IBC §1808.2

The code requires that individual pile caps of every structure subject to seismic forces be interconnected with ties. This is specified in §1808.2.23.1. The ties must be capable of resisting in tension and compression a minimum horizontal tie force equal to 10 percent of the larger column vertical load. The column vertical load is to be considered the dead, reduced live, and seismic loads on the pile cap. An exception to §1808.2.23.1 allows use of "equivalent restraint" which, in this example, is provided by the confinement of very dense granular soil at the site.

2. Interconnection force between pile caps 3 and 10

Maximum loads on each pile cap under E/W seismic forces are

Pile cap 3 = 46 + 16 + 0 = 62 kips

Pile cap 10 = 58 + 16 + 0 = 74 kips

Minimum horizontal tie force $S_{DS}/10 = 0.10$ times the largest column vertical load

P = 0.10 (74) = 7.40 kips

3. Required "tie" restraint between pile caps 3 and 10

The choices are to add a grade beam (i.e., tie beam) connecting pile caps 3 and 10, or to try to use passive pressure restraint on the pile cap in lieu of a grade beam. The latter is considered an "equivalent restraint" (by soil confinement or bearing pressure) under the exception to IBC §1808.2.23.1.

For the allowable lateral bearing = 200 psf per foot of depth below natural grade, the passive pressure resistance is

$$\text{Passive pressure} = \frac{[2(200) + 4(200)]}{2} (2 \text{ ft}) = 1200 \text{ plf}$$

$$\text{Required length} = \frac{7400 \text{ lbs}}{1200 \text{ plf}} = 6.2 \text{ ft}$$

This is greater than 3'-0" pile cap width, but pile cap and a tributary length of N/S grade beam on either side of the pile cap may be designed to resist tie forces using the passive pressure. This system is shown below and, if this is properly designed, no grade beam between pile caps 3 and 10 (or similar caps) is required.

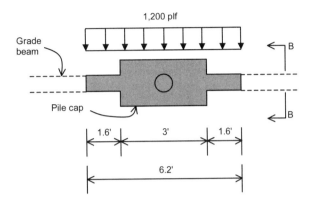

Equivalent restraint system in plan

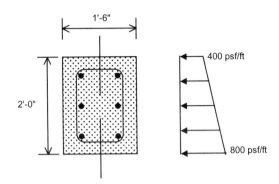

Section B-B: Grade beam

Commentary

Normally, buildings on pile foundations are required to have interconnecting ties between pile caps. This is particularly true in the case of high-rise buildings and buildings with heavy vertical loads on individual pile caps. Ties are essential in tall buildings. Ties are also necessary when the site soil conditions are so poor that lateral movements, or geotechnical hazards, such as liquefaction, are possible.

In the design of relatively lightweight one- and two-story buildings, the exception to the interconnecting tie requirement of §1808.2.23.1 may permit a more economical foundation design. However, when interconnecting ties are omitted, a geotechnical engineer should confirm the appropriateness of this decision, and the project specifications should call for the back-fill and compaction methods necessary to provide required passive pressure resistance.

Example 55 ■ Simplified Wind Loads on 2-Story Buildings   §6.4

Example 55

Simplified Wind Loads on 2-Story Buildings   §6.4

The following is an example of the simplified wind load procedure of ASCE/SEI 7-05.

Calculate the wind loads on the following building.

Dimensions: 100 ft wide by 120 ft long by 25 ft high (2 stories – 13 ft and 12 ft).

Wind Speed: Located in Minneapolis, Minnesota – **90 mph zone**.   F 6-1

Importance: The facility is an office building with no special functions – Therefore the
building category in Table 1-1 is **Category II**.

Exposure: Suburban office park surrounded by trees and typical suburban construction
on all sides – Therefore the exposure category is **B**.   §6.5.6

Enclosure: The building has no unusual openings in the envelope, nor is it in a hurricane
region, so no concerns for wind-borne debris – Classify as **Enclosed**.   §6-2

Topography: Height of adjacent hills is less than 60 feet – Wind speed-up effects not a concern.
(§6.5.7.1.5) $K_{zt} = 1.0$

Structure: The structure is an X-braced steel frame with evenly distributed braces on all four
exterior walls. The second floor is concrete slab on metal form deck on steel floor
beams. The roof is metal roof deck on steel joists on steel joist girders.

Design Method:
To utilize ASCE/SEI 7-05 Simplified Procedure (Method 1) all of the following
criteria must be met.
1) With no breaks in the roof or floor (structural separations) the diaphragms are
simple, as defined in §6-2
2) The building height is less than 60 feet and least horizontal dimensions
3) The building is enclosed and not prone to wind-borne debris
4) The building is regular shaped
5) The building is rigid with a period less than 1 second
6) The site is not subject to wind speed-up effects
7) The building is symmetrical
8) For a building with well distributed MWFRS torsional load case in note 5 of
Figure 6-10 will not govern the design. **Therefore design by Method 1 §6.4**

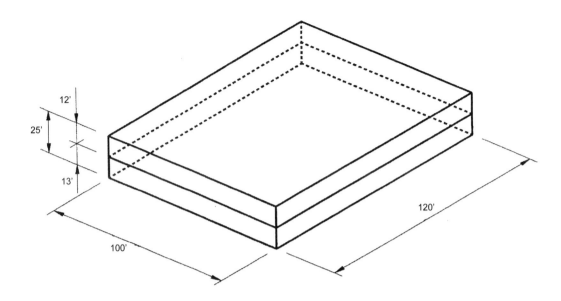

Determine the following.

1.	Main wind force
2.	MWFRS end zone width
3.	MWFRS design wind pressures
4.	Components and cladding
5.	Edge Strip
6.	Design wind pressure on components

Calculations and Discussions   **Code Reference**

1. **Main Wind-Force Resisting System-MWFRS (Lateral Load Structural Frame)**

Using Method 1 §6.4, the simplified design wind pressure p_s is the product of the base simplified design pressure p_{s30} taken from Figure 6-2 and multiplied by the Height and Exposure Adjustment Factor λ from Figure 6-2, the Topographic Factor K_{zt} from §6.5.7, and by the Importance Factor I from Table 6-1. The equation for p_s is shown in §6.4.2.1 Eq 6-1.

2. **Calculate the MWFRS End Zone Width**

End Zone = 2a, so first calculate a, the Edge Strip Width.

Example 55 ■ *Simplified Wind Loads on 2-Story Buildings* **§6.4**

Edge Strip = a = Lesser of:
- 10% of the least horizontal dimension = 0.10 × 100 ft = 10 ft
- 40% of the eave height = 0.40 × 25 ft = 10 ft

But not less than:
- 4% of the least horizontal dimension = 0.04 × 100 ft = 4 ft
- 3 ft

Therefore: a = 10 ft, so the End Zone = 2a = 2 × 10 ft = 20 ft

3. Calculate the MWFRS design wind pressure

Using Equation 6-1: $p_s = \lambda K_{zt} I p_{s30}$
Look up the base pressures p_{s30} from Figure 6-2 then modify for height, exposure, topography, and importance factor. No interpolation is required because the flat roof angle falls in the row of "0 to 5." With the mean roof height of 25 feet and the exposure being "B", the Height and Exposure Adjustment Factor λ from Figure 6-2 = 1.0. Since the building site is level from §6.5.7, K_{zt} = 1.0. For a building Category II as defined in Table 1-1, the Importance Factor I = 1.0.

Transverse MWFRS – 90 mph, Exposure B, Height 25.0 ft

Type	Zone	Surface	Label	p_{s30} Roof Angle 0° to 5°	λ Ht. & Exp. Factor	K_{zt} Topographic Factor	I Import. Factor	p_s Design Pressure
Horiz	End	Wall	A	12.8	λ 1.00	λ 1.00	λ 1.00	= 12.8 psf
		Roof	B		No Roof Projection for Flat Roofs			
	Int	Wall	C	8.5	λ 1.00	λ 1.00	λ 1.00	= 8.5 psf
		Roof	D		No Roof Projection for Flat Roofs			
Vert	End	Wind	E	-15.4	λ 1.00	λ 1.00	λ 1.00	= -15.4 psf
		Lee	F	-8.8	λ 1.00	λ 1.00	λ 1.00	= -8.8 psf
	Int	Wind	G	-10.7	λ 1.00	λ 1.00	λ 1.00	= -10.7 psf
		Lee	H	-6.8	λ 1.00	λ 1.00	λ 1.00	= -6.8 psf

Longitudinal MWFRS – 90 mph, Exposure B, Height 25.0 ft

Type	Zone	Surface	Label	p_{s30} Base Press.	λ Ht. & Exp. Factor	K_{zt} Topographic Factor	I Import. Factor	p_s Design Pressure
Horiz	End	Wall	A	12.8	λ 1.00	λ 1.00	λ 1.00	= 12.8 psf
		Roof	B		No Roof Projection in Longitudinal Direction			
	Int	Wall	C	8.5	λ 1.00	λ 1.00	λ 1.00	= 8.5 psf
		Roof	D		No Roof Projection in Longitudinal Direction			
Vert	End	Wind	E	-15.4	λ 1.00	λ 1.00	λ 1.00	= -15.4 psf
		Lee	F	-8.8	λ 1.00	λ 1.00	λ 1.00	= -8.8 psf
	Int	Wind	G	-10.7	λ 1.00	λ 1.00	λ 1.00	= -10.7 psf
		Lee	H	-6.8	λ 1.00	λ 1.00	λ 1.00	= -6.8 psf

Apply the pressures to the building as described in Figure 6-2. The designations of "Transverse" and "Longitudinal" are keyed to the direction of the MWFRS being evaluated. When the resisting system being designed is perpendicular to the ridge line of the gable or hip roof, its direction is classified as "Transverse." When it is parallel to the ridge, it is classified as "Longitudinal." When the roof is flat (slope ≤5°), and thus has no ridge line, the loading diagram becomes the same in each direction, as shown in the following diagram. The loading diagrams shown should be mirrored about each axis of the building until each of the four corners has been the "reference corner" as shown for each load case.

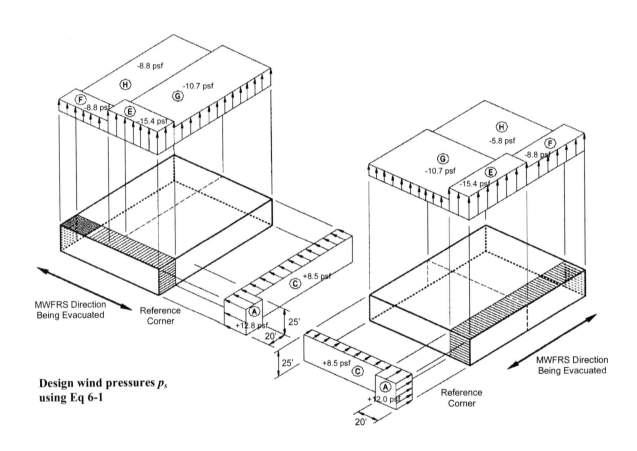

In addition, the minimum load case from §6.4.2.1.1 must also be checked. Apply a load of 10 psf on the building projection on a vertical plane normal to the wind. In other words, create a load case with all horizontal zones equal to 10 psf, and all vertical zones equal to 0. Check this load case as an independent case, do not combine with the case from §6.4.2.1. It should be applied in each direction as well.

Example 55 ■ *Simplified Wind Loads on 2-Story Buildings*   *§6.4*

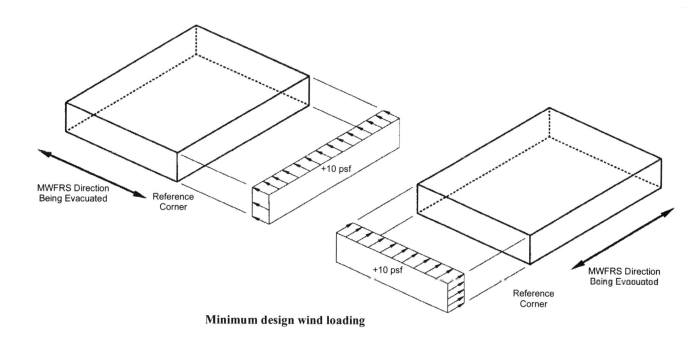

Minimum design wind loading

---

4. Components and Cladding (Everything except the Lateral Load Structural Frame)

According to §6.1.1, all "buildings....and all components and cladding" must be designed for wind loads. Therefore, all parts of the exterior building envelope and any load paths, that are not part of the main wind-force-resisting system (lateral frame), should be designed as Components and Cladding (C&C). For buildings such as this that qualify under §6.4.2.1, the C&C can be designed using §6.4.2.2, Eq 6-2.

---

5. Calculate the Edge Strip, a

Previously calculated in the MWFRS calculations, a = 10 ft

6. Calculate the design wind pressure on several components using Equation 6-2

$$p_{net} = \lambda \ K_{zt} I p_{net30}$$

Look up the base pressures directly from Figure 6-3, then modify for Height, Exposure, Topography and Importance Category. With the mean roof height of 25 feet and the exposure being "B," the Height and Exposure Adjustment Factor from Figure 6-3 = 1.00. Since the building is a level site from §6.5.7, $K_{zt} = 1.0$, and for a Building Category II, the Importance Factor $I_w = 1.00$.

C & C – 90 mph, Exposure B, Height = 25.0 ft

Type	Zone	Item	Eff Wind Area	Direction	Interpolation			p_{net30} Base Press	λ Ht. & Exp. Factor	K_{zt} Topo. Factor	I Import. Factor	p_{s30} Design Pressure
Roof 0° to 7°	Int (1)	Deck Screw	< 10 sf	Positive	None Required			+5.9	× 1.00	× 1.00	× 1.00	+5.9*
				Negative	None Required			-14.6	× 1.00	× 1.00	× 1.00	-14.6
		Roof Deck	12 sf	Positive	10 sf / +5.9	20 sf / +5.6	12 sf / +5.8	+5.8	× 1.00	× 1.00	× 1.00	+5.8*
				Negative	10sf / −14.6	20 sf / −14.2	12 sf / −14.5	-14.5	× 1.00	× 1.00	× 1.00	-14.5
		Joist	> 100 sf	Positive	None Required			+4.7	× 1.00	× 1.00	× 1.00	+4.7*
				Negative	None Required			-13.3	× 1.00	× 1.00	× 1.00	-13.3
	Edge (2)	Deck Screw	< 10 sf	Positive	None Required			+5.9	× 1.00	× 1.00	× 1.00	+5.9*
				Negative	None Required			-24.4	× 1.00	× 1.00	× 1.00	-24.4
		Roof Deck	12 sf	Positive	None Required			+5.8	× 1.00	× 1.00	× 1.00	+5.8*
				Negative	10 sf / -24.4	20 sf / -21.8	12 sf / -23.9	-23.9	× 1.00	× 1.00	× 1.00	-23.9
		Joist	> 100 sf	Positive	None Required			+4.7	× 1.00	× 1.00	× 1.00	+4.7*
				Negative	None Required			-15.8	× 1.00	× 1.00	× 1.00	-15.8
	Corner (3)	Deck Screw	< 10 sf	Positive	None Required			+5.9	× 1.00	× 1.00	× 1.00	+5.9*
				Negative	None Required			-36.8	× 1.00	× 1.00	× 1.00	-36.8
		Roof Deck	12 sf	Positive	None Required			+5.8	× 1.00	× 1.00	× 1.00	+5.8*
				Negative	10 sf / -36.8	20 sf / -30.5	12 sf / -35.5	-35.5	× 1.00	× 1.00	× 1.00	-35.5
		Joist	> 100 sf	Positive	None Required			+4.7	× 1.00	× 1.00	× 1.00	+4.7*
				Negative	None Required			-15.8	× 1.00	× 1.00	× 1.00	-15.8
Wall	Int (4)	Siding	< 10 sf	Positive	None Required			14.6	× 1.00	× 1.00	× 1.00	14.6
				Negative	None Required			-15.8	× 1.00	× 1.00	× 1.00	-15.8
		Stud	17.3 sf	Positive	10 sf / +14.6	20 sf / +13.9	17.3 sf / +14.1	+14.1	× 1.00	× 1.00	× 1.00	+14.1
				Negative	10 sf / -15.8	20 sf / -15.1	17.3 sf / -15.3	-15.3	× 1.00	× 1.00	× 1.00	-15.3
	Int (4)	Siding	< 10 sf	Positive	None Required			+14.6	× 1.00	× 1.00	× 1.00	+14.6
				Negative	None Required			-19.5	× 1.00	× 1.00	× 1.00	-19.5
		Stud	17.3 sf	Positive	10 sf / +14.6	20 sf / +13.9	17.3 sf / +14.1	+14.1	× 1.00	× 1.00	× 1.00	+14.1
				Negative	10 sf / -19.5	20 sf / -18.2	17.3 sf / -18.6	-18.6	× 1.00	× 1.00	× 1.00	-18.6

* Note: A minimum pressure of 10 psf is required per §6.4.2.2.1

Example 55 ■ *Simplified Wind Loads on 2-Story Buildings* §*6.4*

The component and cladding pressures should be applied as described in Figure 6-3 and as shown in the diagram below.

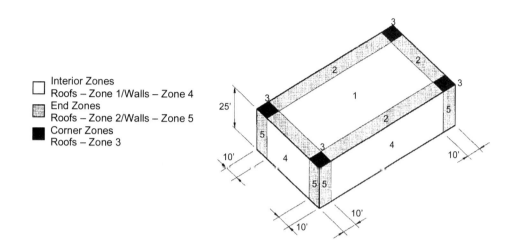

Example 56
Simplified Wind Loads on Low-Rise Buildings §6.4

Per §6.4.1.1, for conforming low-rise buildings, wind loads can be determined using simplified provisions.

The following information is given.

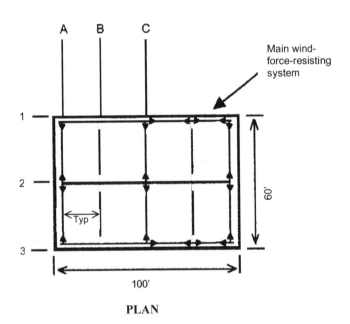

PLAN

3-story office building located in urban/suburban area of NW Texas – situated on flat ground

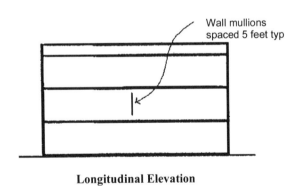

Longitudinal Elevation

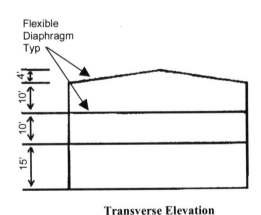

Transverse Elevation

Determine the following.

1. **Wind loads on MWFRS at Grid A**

2. **Wind loads on second-story wall mullion**

Example 56 ■ *Simplified Wind Loads on Low Rise Buildings*   *§6.4*

Calculations and Discussion	*Code Reference*

1. Wind loads on MWFRS at Grid A

1a. Check applicability of simplified provisions §6.4.1.1

1. Simple diaphragm building (See definition under "building, simple diaphragm")	Yes	§6.2
2. Low rise building (Mean roof height < 60 ft and building width)	Yes	§6.2
3. Building enclosed	Yes	§6.2
4. Regular shape	Yes	§6.2
5. Not flexible ($n_1 > 1$ hz) ($T < 1$ sec)	Yes	§6.2

$$T = 0.1 \, N = 0.1(3) = 0.3 \text{ sec}$$
$$N = \text{Number of Stories}$$

6. No special wind characteristics	Yes	
7. Flat, gabled or hipped roof	Yes	
8. Torsional irregularities not a concern	Yes	Note 5, F 6-10

Therefore, simplified provisions are applicable

1b. Determine basic parameters

NW Texas basic wind speed = 90 mph   F 6-1a
The design professional should contact the local building department to
confirm design wind speed.

Height and exposure adjustment λ F 6-2
See §6.5.6 for exposure category definitions

Example building in urban/suburban area is considered exposure B
Mean roof height (h) = 35 ft (see definition §6.2) ($\theta < 10°$)
Adjustment factor from Figure 6-2, $\theta = 1.05$

Topographic factor $K_{zt} = 1.0$ §6.5.7

Importance Factor $I = 1.0$ T 6-1
(Category II Building from Table 1-1)

1c. **Obtain tabulated loads**   **F 6-2**

Simplified Design Wind Pressure p_{s30} (psf)

V	Load Dir.	Roof Angle	Horizontal Loads				Vertical Loads			
			End Zone		Int. Zone		End Zone		Int. Zone	
			A Wall	B Roof	C Wall	D Roof	E WW Roof	F LW Roof	G WW Roof	H LW Roof
90 mph	Transverse	0 to 5°	12.8	–6.7	8.5	–4.0	–15.4	–8.8	–10.7	–6.8
			17.8	–4.7	11.9	–2.6	–15.4	–10.7	–10.7	–8.1
		20°								
Interpolating: For example, roof angle $= \arctan \frac{4}{30} = 7.6°$		7.6°	13.7	–6.4 (use 0)	9.1	–3.8 (use 0)	–15.4	–9.1	–10.7	–7.0

1d. **Determine end zone dimensions**   **Note 10, F 6-2**

Edge Strip   a $= 0.10\,(60) = 6$ ft ... Governs
               or
                  $= 0.40\,(35) = 14$ ft
               but not less than
                  $\geq 0.04\,(60) = 2.4$ ft
               or
                  ≥ 3 ft

End Zone   2a $= 12$ ft   F 6-2

1e. **Determine load on MWFRS at Grid A**   **§6.4.2.1**

Forces determined using Eq 6-1 $p_s = \lambda\, K_{zt}\, I\, p_{s30}$

Horizontal load at wall:
    In end zone $[A] = (1.05)(1.0)(1.0)(13.7 \text{ psf}) = 14.4$ psf
    In interior zone $[C] = (1.05)(1.0)(1.0)(9.1 \text{ psf}) = 9.6$ psf

Per §6.1.4.1, check 10 psf minimum over projected area of vertical plane

Check minimum requirement:
    Horizontal load Eq 6-1 $= (14.4 \text{ psf}*12 \text{ ft} + 9.6 \text{ psf}*(25-12))*35 \text{ ft} = 10.42$ kips
    Min load §6.1.4.1 $= (10 \text{ psf}* 25 \text{ ft})*35 \text{ ft} = 8.75$ kips < 10.42
                                         $\therefore$ 6.1.4.1 does not govern

Example 56 ■ *Simplified Wind Loads on Low Rise Buildings* §6.4

Horizontal point loads to frame:

Roof Load

(5 ft tributary ht) $V_R = (14.4 \text{ psf}*12 \text{ ft} + 9.6 \text{ psf}*(25 \text{ ft} - 2 \text{ ft}))5 \text{ ft} = 1488 \text{ lb}$

3rd Floor Load

(10 ft tributary ht) $V_3 = (14.4 \text{ psf}*12 \text{ ft} + 9.6 \text{ psf}*(25 \text{ ft} - 2 \text{ ft}))10 \text{ ft} = 2976 \text{ lb}$

2nd Floor Load

(12.5 ft tributary ht) $V_2 = (14.4 \text{ psf}*12 \text{ ft} + 9.6 \text{ psf}*(25 \text{ ft} - 2 \text{ ft}))12.5 \text{ ft} = 3720 \text{ lb}$

Note: Forces to Grid A are shown based on a tributary basis that is conservative for the analysis of Grid A. Alternatively, the forces could be distributed to grids A and C by applying the loads as a simple span beam.

Vertical load at roof:

Windward Roof – In end zone $[E] = (1.05)(1.0)(1.0)(-15.4 \text{ psf}) = -16.2 \text{ psf}$

                In interior zone $[G] = (1.05)(1.0)(1.0)(-10.7 \text{ psf}) = -11.2 \text{ psf}$

Leeward Roof –   In end zone $[F] = (1.05)(1.0)(1.0)(-9.1 \text{ psf}) = -9.56 \text{ psf}$

                In interior zone $[H] = (1.05)(1.0)(1.0)(-7.0 \text{ psf}) = -7.35 \text{ psf}$

Vertical uniform loads to frame:

Windward: $(16.2 \text{ psf})(12 \text{ ft}) + (9.56 \text{ psf})(25 - 12) = 340 \text{ plf} = .34 \text{ klf uplift}$

Leeward: $(11.2 \text{ psf})(12 \text{ ft}) + (7.35 \text{ psf})(25 - 12) = 210 \text{ plf} = .21 \text{ klf uplift}$

Note: Forces applied to Grid A are shown as a distributed load along the frame length. A more detailed analysis of forces based on roof framing would include a smaller distributed load and uplift point loads at locations where beams frame into the grid A moment frame at grids 1, 2, and 3.

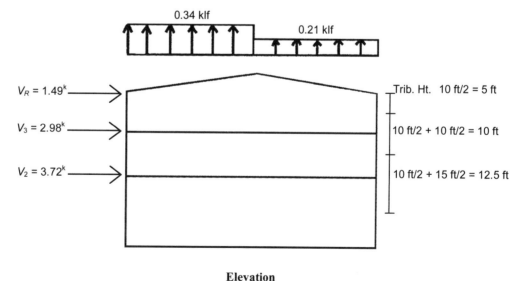

Elevation

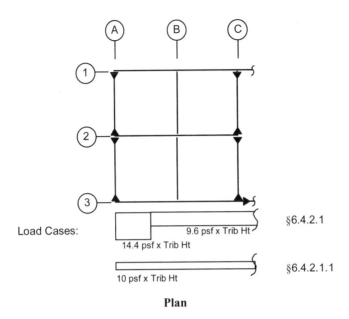

Plan

---

2. Wind loads on second story wall mullion

2a. Determine zone of mullion   F 6-3

Interior of wall area – Zone 4

Effective wind area = 5 ft (10 ft) = 50 sq ft

Wind Loads $p_s = \lambda\, K_{zt} I\, p_{net30}$ §6.4.2.2 (Eq 6-2)

p_{net30} = 13.0 psf positive
           = −14.3 psf negative (suction)   F 6-3

p_s = (1.05)(1.0)(1.0)(13.0 psf$_{positive}$)(5 ft tributary) = 68.5 plf
p_s = (1.05)(1.0)(1.0)(−14.3 psf$_{negative}$)(5 ft tributary) = 75 plf

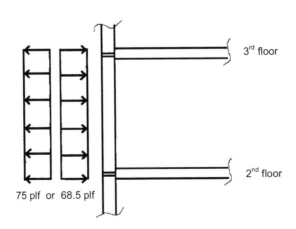

Example 58 ■ *Floor Vibrations*

Example 57

Wind Loads – Analytical Procedure

§6.5

A 9-story building has a moment-resisting frame for a lateral force-resisting system.

Find the lateral forces on the frame due to wind.

Office building 50 ft by 50 ft in
plan with MWFRS at exterior.
Located in an urban/suburban
area of N.W. Texas

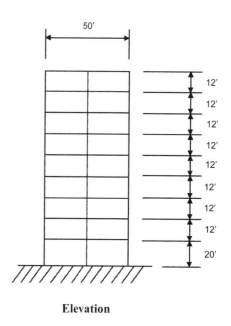

Elevation

Determine:

1. **Wind loads on MWFRS**

Calculations and Discussion

1. **Wind loads on MWFRS** **Chapter 6**

1a. **Determine basic wind speed**

Utilize ASCE/SEI 7-05 §6

Use method 2 analytical procedure §6.5

Confirm building is regular shaped and not subject to across wind loading, vortex shedding, instability due to galloping or flutter; or does not have a site location for which channeling effects or buffeting in wake of upwind obstructions warrant special conditions

§6.5.1

Design procedure

§6.5.3

Basic wind speed $V = 90$ mph

§6.5.4, F 6-1

1b. Determine velocity pressure

Wind directionality factor $K_d = 0.85$
  (applies when using load combinations
  in ASCE/SEI 7-05 §2.3 and §2.4)

§6.5.4.4, T 6-4

Importance factor $I = 1.00$
  (Structural Category II, Table 1-1)

§6.5.5, T 6-1

Exposure Category B

§6.5.6

Velocity pressure coeff K_z (Case 2)

§6.5.6.6, T 6-3

h	Exposure B Case 2
0-15 ft	0.57
20	0.62
25	0.66
30	0.70
40	0.76
50	0.81
60	0.85
70	0.89
80	0.93
90	0.96
100	0.99
116	1.03 ◄—— By Interpolation
120	1.04

Topographic factor $K_{Zt} = 1$
  (example building on flat land, no nearby hills)

§6.5.7

Gust effect factor G
  9-story building
  Natural period $= 0.1(9) = 0.9$ sec

§6.5.8

§12.8.2.1

  Natural frequency $= \dfrac{1}{0.9} = 1.1$ Hz >1.0

(Eq 12.8-7)

Therefore: Rigid structure
  $G = 0.85$

§6.2
§6.5.8.1

Example 58 ■ *Floor Vibrations*

Enclosure Classification §6.5.9
    Example building enclosed

Velocity Pressure §6.5.10
$$q_z = 0.00256K_2K_{2k}KdV^2I$$ Eq 6-15
$$= 0.00256K_zK_{zt}K_2V^2I$$
$$= 0.00256K_z(1.0)(0.85)(90)^2(1.0)$$

h	q_z
0-15 ft	10.0 psf
20	10.9
25	11.6
30	12.3
40	13.4
50	14.3
60	15.0
70	15.7
80	16.4
90	16.9
100	17.4
116	18.2

1c. Determine pressure and force coefficients §6.5.11

Internal pressure coefficients - GC_{pi} §6.5.11.1,
    $GC_{pi} = \pm0.18$ Case 1: Internal Pressure Inward F 6-5
                   Case 2: Internal Pressure Outward

External pressure coefficients - C_p §6.5.11.2,
    For example building, monoslope roof $\theta = 0$ F 6-6

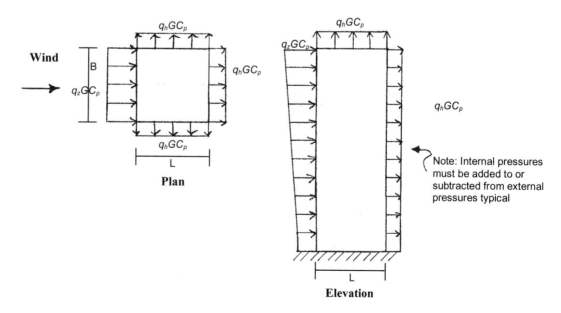

Note: Internal pressures must be added to or subtracted from external pressures typical

Windward wall $C_p = 0.8$ F 6-6

h	$q_z GC_p = q_z (0.85)(0.8)$
0-15 ft	6.80
20	7.41
25	7.89
30	8.36
40	9.11
50	9.72
60	10.2
70	10.7
80	11.2
90	11.5
100	11.8
116	12.4

Leeward wall

$$\frac{L}{B} = \frac{50}{50} = 1 \rightarrow C_p = -0.5$$ F 6-6

$$q_h = q_{h=116\,ft} = 18.2 \text{ psf}$$

$$q_h GC_p = 18.2 (0.85)(-0.5) = -7.74 \text{ psf}$$

Side walls

$$C_p = -0.7$$ F 6-6

$$q_h GC_p = 18.2 (0.85)(-0.7) = -10.8 \text{ psf}$$

Roof

$$\frac{h}{L} = \frac{116}{50} = 2.3 > 1.0$$

$$C_p = -1.3 \times 0.8 \text{ (Area Reduction Factor)} = 1.04$$ F 6-6

$$q_h GC_p = 18.2 \text{ psf}(0.85)(\times1.04) = \times16.1 \text{ psf}$$

1d. **Design wind loads** **§6.5.12**

Main wind-force-resisting system §6.5.12.2

Rigid building §6.5.12.2.1

Example 58 ■ *Floor Vibrations*

$$p = qGC_p - q_i(GC_{pi})$$ (Eq 6-17)

Windward wall

$$q_h(GC_{pi}) = (18.2)(0.18) = 3.28 \text{ psf}(\pm)$$

h	$p = q_z GC_p - q_h(GC_{pi})$	Case 1 shown
0-15 ft	10.1	
20	10.7	
25	11.2	
30	11.6	
40	12.4	
50	13.0	
60	13.5	
70	14.0	
80	14.5	
90	14.8	
100	15.1	
116	15.7	

Sample Calculation
$p = 12.4 - 18.2(-0.18) = 15.7$ Case 1
$12.4 - 18.2(+0.18) = 9.1$ Case 2

Leeward wall

$$p = q_h GC_p - q_h (GC_{pi})$$

$$p = -7.74 - 18.2(-0.18) = -4.5 \text{ psf} \quad \text{Case 1}$$

$$p = -7.74 - 18.2(0.18) = -11.0 \text{ psf} \quad \text{Case 2}$$

Side walls

$$p = q_h GC_p - q_h (GC_{pi})$$

$$= -10.8 - 18.2(0.18) = -14.1 \text{ psf}$$

Roof

$$p = q_h GC_p - q_h(GC_{pi})$$

$$= -16.1 - 18.2(0.18) = -19.4 \text{ psf}$$

1e. Design wind loads – graphically

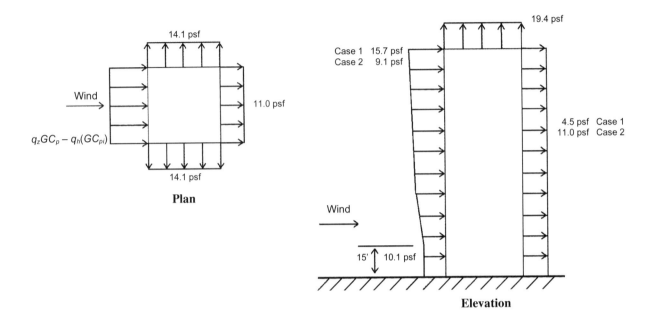

Plan

Elevation

Case 1: Internal Pressure Inward
Case 2: Internal Pressure Outward

Verify projected load is greater than 10 psf   §6.1.4.1
$10.1 + 11.0 = 21.1 > 10$ psf . . .*o.k.*

To obtain frame loads, multiply pressures by tributary width = $50/2 = 25$ ft or perform Rigid Diaphragm Analysis